Learning Android Game Development

Android Game development simplified!

Nikhil Malankar

BIRMINGHAM - MUMBAI

Learning Android Game Development

First published: May 2017

Production reference: 1240517

Published by Packt Publishing Ltd.
Livery Place
35 Livery Street
Birmingham
B3 2PB, UK.
ISBN 978-1-78588-095-7

www.packtpub.com

Credits

Author
Nikhil Malankar

Reviewer
Nischal Dubey

Commissioning Editor
Amarabha Banerjee

Acquisition Editor
Shweta Pant

Content Development Editor
Aditi Gour

Technical Editor
Rashil Shah

Copy Editor
Dhanya Baburaj

Project Coordinator
Ritika Manoj

Proofreader
Safis Editing

Indexer
Rekha Nair

Graphics
Jason Monteiro

Production Coordinator
Nilesh Mohite

About the Author

Nikhil Malankar started his journey into game development in 2011 by founding his company, GameEon, as the CEO, at the age of 17. GameEon has developed over 70+ games, of which 10 are available for download on Play Store and others are distributed worldwide via multiple distribution channels. Currently, he is running his new company--Next Move Digital--as the founder and CEO. Next Move Digital deals in digital media and game development.

He has a total experience of 5 years in the field of game development and has worked on technologies such as Pygame (a subset of Python) to create GameEon's first game, Kyte - Kite Flying Game, which has over 300,000 downloads on Google Play. He is also familiar with the Unity Game engine and has developed most of the games at GameEon in the same. He has also worked with Unreal Engine 4 to develop Special Ops, a first person shooter game for Android and iOS.

He has also developed non-gaming apps and websites for clients. At GameEon, he also worked with multiple clients to develop games for them, and one of the most famous brands he has worked with is m-Indicator. He is also a social media influencer with a big following of his own on Facebook and is extremely passionate about playing games. With Next Move Digital, he aims to work on content creation, distribution, and licensing. He also operates a content website--Tell Me Nothing--under Next Move Digital, which focuses on lighthearted satirical content. His future endeavors with the company include making good-quality games for PC and console platforms.

Currently, he is running Next Move Digital, as the Founder and CEO, that focuses on creating digital media content.

I would like to thank Shweta Pant for introducing me to Packt and giving me the opportunity to work on this book. Also, I would like to thank Aditi Gour, who has been extremely supportive and friendly throughout the entire process of writing the book. I would like to thank Rashil Shah, the technical editor on the book, and I would also like to thank Raimon Rafols and Nischal Dubey, the technical reviewers of this book, for ensuring that the information presented by me was technically accurate.
Of course, I would also thank my mom, dad, and sister for supporting me throughout the entire process of writing this book, and also my close friends, without whom this book wouldn't have been possible.

About the Reviewer

Nischal Dubey is a full stack application developer who specializes in Java development. He has more than 3 years of experience in android development and has made applications having features ranging from basic utility apps to complex social networking applications. Over the years, he has tried to master in and out of Android development. Currently, he is working as Java and Angular JS developer in TCS.

www.PacktPub.com

For support files and downloads related to your book, please visit www.PacktPub.com.

Did you know that Packt offers eBook versions of every book published, with PDF and ePub files available? You can upgrade to the eBook version at www.PacktPub.com and as a print book customer, you are entitled to a discount on the eBook copy. Get in touch with us at service@packtpub.com for more details.

At www.PacktPub.com, you can also read a collection of free technical articles, sign up for a range of free newsletters and receive exclusive discounts and offers on Packt books and eBooks.

https://www.packtpub.com/mapt

Get the most in-demand software skills with Mapt. Mapt gives you full access to all Packt books and video courses, as well as industry-leading tools to help you plan your personal development and advance your career.

Why subscribe?

- Fully searchable across every book published by Packt
- Copy and paste, print, and bookmark content
- On demand and accessible via a web browser

Customer Feedback

Thanks for purchasing this Packt book. At Packt, quality is at the heart of our editorial process. To help us improve, please leave us an honest review on this book's Amazon page at `https://www.amazon.com/dp/1785880950`.

If you'd like to join our team of regular reviewers, you can e-mail us at `customerreviews@packtpub.com`. We award our regular reviewers with free eBooks and videos in exchange for their valuable feedback. Help us be relentless in improving our products!

Table of Contents

Preface

This book will take our readers to a fun-filled ride where they will not just learn about the latest Android N SDK, but also about the other APIs and how they can create highly interactive and interesting games with them. The book will show readers how to create a complete game from scratch, designed for the Android platform. It will start by setting up the Android N SDK and other repositories, and then it will show readers how to customize the development environment. After this, it will show readers how to create game elements, objects, game layouts, game architecture, and game loops. It will create reusable Java code scripts, which will help you in your other game projects. An integral part of game development is to integrate images and graphics seamlessly. As we move ahead, we will show how to efficiently work with moving images, creating sprite animations, particle explosions, in-game entities, Bitmap fonts, and so on. Prototyping can decrease development time significantly; readers will implement prototyping techniques using the libgdx library. Toward the end of the book, readers will get a clear idea of improving game physics and the collision system, to give their game that real-life look.

What this book covers

Chapter 1, *Introduction to Android N and Installation of Android SDK*, guides the reader through the installation of the necessary software with a step-by-step guide.

Chapter 2, *Getting Familiar with Android Studio*, works toward making the reader comfortable with the project layout and components needed to get started with developing games with Android Studio.

Chapter 3, *Managing Inputs*, guides the reader through how to take inputs from the user.

Chapter 4, *Creating Sprites and Interactive Objects*, teaches how to display images on screen and turn them into interactive objects.

Chapter 5, *Adding Animation to Your Game*, takes you through how to create animations using sprite sheets.

Chapter 6, *Collision Detection and Basic Artificial Intelligence*, explores various collision detection methods. It is recommended that the user has a little mathematical understanding of some basic concepts of the coordinates system.

Chapter 7, *Adding Boundaries and Using Sprites to Create Explosions*, goes through creating boundaries for the game, and then moves on to create an explosion class to create an explosion effect.

Chapter 8, *Adding an Explosion and Creating a UI*, shows how we will spawn our explosion on screen after creating the explosion class, and then covers on how to create a UI that will display instructions and score the on screen.

Chapter 9, *Converting Your Game from 2D to 3D*, discusses how to move from making 2D games into 3D games.

Chapter 10, *Working Further on 3D Game*, introduces the player to further work with 3D objects in the game.

What you need for this book

This book is aimed at those who want to get started with native development with Android. It also gives a ready introduction to the latest version of Android, Android N, and guides the reader through the process of going from developing a simple app to a complex 3D game in a matter of 10 simple chapters.

Who this book is for

This book is for anyone who has a basic knowledge of Java and is interested in making games for Android. No prior knowledge of Android or game development is needed; however, this would be a plus.

Conventions

In this book, you will find a number of text styles that distinguish between different kinds of information. Here are some examples of these styles and an explanation of their meaning. Code words in text, database table names, folder names, filenames, file extensions, pathnames, dummy URLs, user input, and Twitter handles are shown as follows: "For instance, you can think of `www.google.com` as a website's domain name." A block of code is set as follows:

```
<TextView
    android:text="Hello World!"
    android:layout_width="wrap_content"
    android:layout_height="wrap_content"
    android:id="@+id/TextView"
    tools:text="helloWorld"
    android:textAppearance="@style/TextAppearance.AppCompat.Headline" />
```

New terms and **important words** are shown in bold. Words that you see on the screen, for example, in menus or dialog boxes, appear in the text like this: "Click on **Text**, as shown in the following screenshot:"

Warnings or important notes appear in a box like this.

Tips and tricks appear like this.

Reader feedback

Feedback from our readers is always welcome. Let us know what you think about this book-what you liked or disliked. Reader feedback is important for us as it helps us develop titles that you will really get the most out of. To send us general feedback, simply e-mail `feedback@packtpub.com`, and mention the book's title in the subject of your message. If there is a topic that you have expertise in and you are interested in either writing or contributing to a book, see our author guide at `www.packtpub.com/authors`.

Customer support

Now that you are the proud owner of a Packt book, we have a number of things to help you to get the most from your purchase.

Downloading the example code

You can download the example code files for this book from your account at http://www.p acktpub.com. If you purchased this book elsewhere, you can visit http://www.packtpub.c om/support and register to have the files e-mailed directly to you. You can download the code files by following these steps:

1. Log in or register to our website using your e-mail address and password.
2. Hover the mouse pointer on the **SUPPORT** tab at the top.
3. Click on **Code Downloads & Errata**.
4. Enter the name of the book in the **Search** box.
5. Select the book for which you're looking to download the code files.
6. Choose from the drop-down menu where you purchased this book from.
7. Click on **Code Download**.

Once the file is downloaded, please make sure that you unzip or extract the folder using the latest version of:

- WinRAR / 7-Zip for Windows
- Zipeg / iZip / UnRarX for Mac
- 7-Zip / PeaZip for Linux

The code bundle for the book is also hosted on GitHub at https://github.com/PacktPubl ishing/-Learning-Android-Game-Development. We also have other code bundles from our rich catalog of books and videos available at https://github.com/PacktPublishing/. Check them out!

Errata

Although we have taken every care to ensure the accuracy of our content, mistakes do happen. If you find a mistake in one of our books-maybe a mistake in the text or the code-we would be grateful if you could report this to us. By doing so, you can save other readers from frustration and help us improve subsequent versions of this book. If you find any errata, please report them by visiting http://www.packtpub.com/submit-errata, selecting your book, clicking on the **Errata Submission Form** link, and entering the details of your errata. Once your errata are verified, your submission will be accepted and the errata will be uploaded to our website or added to any list of existing errata under the Errata section of that title. To view the previously submitted errata, go to https://www.packtpub.com/books/content/support and enter the name of the book in the search field. The required information will appear under the **Errata** section.

Piracy

Piracy of copyrighted material on the Internet is an ongoing problem across all media. At Packt, we take the protection of our copyright and licenses very seriously. If you come across any illegal copies of our works in any form on the Internet, please provide us with the location address or website name immediately so that we can pursue a remedy. Please contact us at copyright@packtpub.com with a link to the suspected pirated material. We appreciate your help in protecting our authors and our ability to bring you valuable content.

Questions

If you have a problem with any aspect of this book, you can contact us at questions@packtpub.com, and we will do our best to address the problem.

1
Introduction to Android N and Installation of Android SDK

Welcome to the world of Android and game development. You are about to begin a journey that will set up a foundation for you to get started with converting your wildest imaginations into games. This book will be your stepping stone to creating amazing games. If you are a complete newbie, you will go through a steep yet comfortable learning curve, and, by the end of this book, you will have created your own game.

This book's chapters have been divided into extremely easy-to-understand parts, which require no prior experience in game development. Experience in programming, however, is a must.

This chapter will guide you through an introduction to Android N along with steps for installation of required software. In short, you will be learning the following in this chapter:

- Short introduction to Android N
- Introduction to game development with a few examples of games that are doing well
- Installation of Android Studio
- Components of Android Studio and setting up for Android N
- Quick introduction to some basic concepts in Android

Introduction to Android N

It all started way back in 2005 when Google acquired a new company, which would later change the course of mobile computing for good. Yes, you guessed it right! The company that was acquired was the developer of the Android operating system. Since then, Android has seen a lot of developments and has grown significantly in terms of its user base because of the might of Google. At the time of writing this book, Android N is the latest version of this OS. Market share of Android has been growing ever since and is currently at 87.6% of total mobile computing devices. This is huge, and therefore, from a developer perspective, it is extremely important to develop for this platform because most of the mobile base are Android users.

Android N stands for Android Nougat. You must be aware of the naming convention for Android versions. If you are not, they are named in an incremented alphabetical fashion and each version is named after a sweet, barring the exception of the first two versions. Here's a quick look at the different versions of Android:

- Alpha
- Beta
- Cupcake
- Donut
- Eclair
- Froyo
- Gingerbread
- Honeycomb
- Ice Cream Sandwich
- Jellybean
- Kit Kat
- Lollipop
- Marshmallow
- Nougat

You can read more about the history of android from the official source at https://www.android.com/history/.

The world of app development is interesting-but even more interesting than that is, one specific field, that is, game development. Mobile games account for the highest number of downloads on the Google Play Store, and, therefore, this is a most exciting time for game developers since Google has established a massive distribution channel and has made it extremely easy for mobile game developers to publish their games. Gone are the days when you'd have to wait for months or even years to crack a publishing deal with a major publisher. In today's times, you can simply sign up on Google Play Store as a developer and in a matter of hours publish your first game, if it is ready, and get feedback from live users.

The world of Android games has seen massive success stories, such as Angry Birds, Candy Crush, Subway Surfers, and so on. Even games with simple gameplay, such as Flappy Bird, have done extremely well, and it was estimated that at the game's peak it was earning around $50,000 per day in ad revenues. Isn't that exciting? You are just one click away from getting your game to a potential audience of billions, and, if your game gets noticed, then you'll be having the time of your life.

You can make a game as simple as a text-based game or as complex as a third-person shooter. You are only as restricted as your imagination. Plus, all the resources you need are available easily today online. This book will serve you as a ready reference to get started in the world of Android game development and will use the latest version of Android, so you are up-to-date with your knowledge. You don't necessarily need to have prior experience of developing games for Android platform; however, if you do, then that would be a plus. You do need to have a little bit of Java programming experience, though, to get started. However, rest assured that this book's language is going to be as easy to understand as possible, and in the whole process of developing your first game for Android, you will have a lot of fun.

So, without further ado, let's dive right into this exciting journey of developing games for Android using the latest components and tools available at our disposal. I hope you have a great time reading and implementing simultaneously from this book and would highly recommend that you make your own notes while going through this book.

Software requirements

To start our journey into game development for Android, we will need certain software installed on your computer. We will use the latest version of Android Studio, as of this writing, to get started. This chapter will guide you through the installation process.

Before you start your installation, make sure that your computer meets the following system requirements:

- Windows:
    - Microsoft® Windows® 7/8/10
    - 3 GB RAM minimum, 8 GB RAM recommended; plus 1 GB for the Android Emulator
    - 2 GB of available disk space minimum, 4 GB recommended (500 MB for IDE + 1.5 GB for Android SDK and emulator system image)
    - 1280 x 800 minimum screen resolution
    - For accelerated emulator, 64-bit operating system and Intel® processor with support for Intel® VT-x, Intel® EM64T (Intel® 64), and Execute Disable (XD) Bit functionality

- Mac:
    - Mac® OS X® 10.10 (Yosemite) or higher, up to 10.12 (macOS Sierra)
    - 3 GB RAM minimum, 8 GB RAM recommended; plus 1 GB for the Android Emulator
    - 2 GB of available disk space minimum, 4 GB recommended (500 MB for IDE + 1.5 GB for Android SDK and emulator system image)
    - 1280 x 800 minimum screen resolution

- Linux:
    - GNOME or KDE desktop
    - 64-bit distribution capable of running 32-bit applications
    - GNU C Library (glibc) 2.11 or later
    - 3 GB RAM minimum, 8 GB RAM recommended; plus 1 GB for the Android Emulator
    - 2 GB of available disk space minimum, 4 GB recommended (500 MB for IDE + 1.5 GB for Android SDK and emulator system image)
    - 1280 x 800 minimum screen resolution
    - For accelerated emulator, Intel®processor with support for Intel®VT-x, Intel® EM64T (Intel® 64), and Execute Disable (XD) Bit functionality, or AMD processor with support for AMD Virtualization (AMD-V)

Emulator acceleration requires that you install either **Intel Hardware Accelerated Execution Manager** (**Intel HAXM**) or **Kernel-based Virtual Machine** (**KVM**), which are types of hypervisors. If the needed hypervisor isn't installed, Android Studio typically prompts you to install it. Without acceleration, the emulator takes the machine code from the VM and translates it block-by-block to conform to the architecture of the host computer. This process can be quite slow. However, if the VM and the architecture of the host computer matches (such as x86 on x86), the emulator can skip translating the code and simply run it directly on the actual CPU using a hypervisor. In this case, the emulator can approach the speed of your actual computer.

You can start installing Android Studio from the following URL:

```
https://developer.android.com/studio/index.html
```

For writing of this book, we have used a Windows 10 system with minimum system requirements. Once you have downloaded Android Studio's .exe file, go through the following steps to finish the installation:

1. Open the .exe file that you have just downloaded
2. Follow the setup wizard and install it using **Standard installation**

Once you have done this, you will be ready to launch Android Studio with the SDK components needed for Android N; SDK tools version for Android is 25.0.0.

The installation steps are as follows:

Press **Next** to start with the setup:

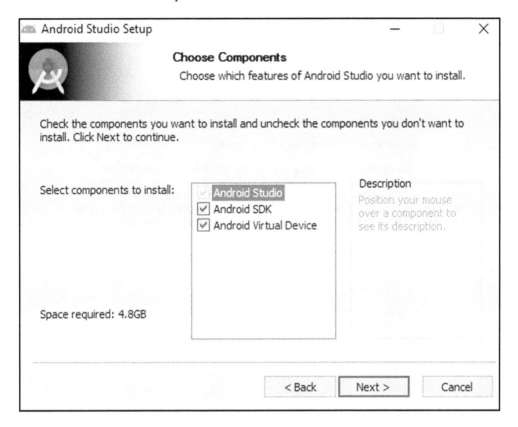

Make sure that you have enough space for installation and then proceed by clicking **Next**:

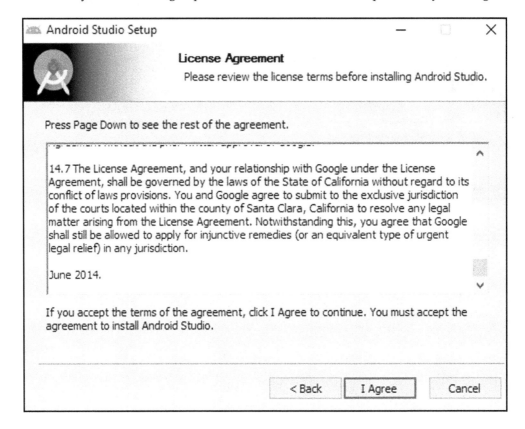

Once you have read through the terms and agreements, press **I Agree** to proceed:

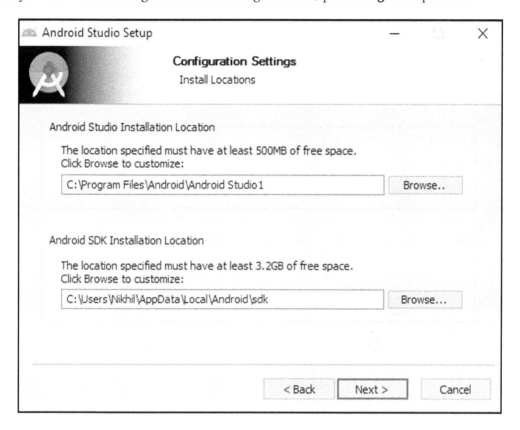

Select your desired path for the installation of Android Studio and press **Next**:

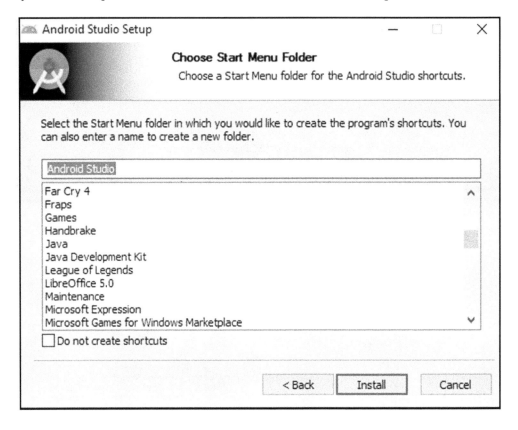

Create a start menu item for quick access and press **Install**:

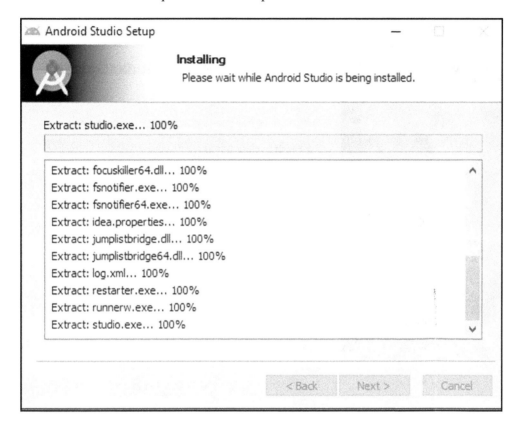

Wait until the installation process finishes:

You are now finished with the installation of Android Studio; press **Finish** to proceed.

Now, you need to configure Android Studio with Android N SDK. The steps to do so, are illustrated as follows:

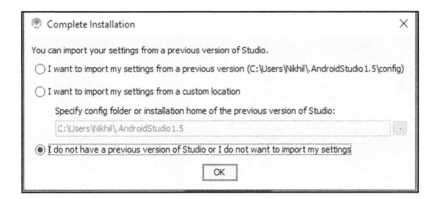

Since we are installing a fresh copy, select the last option as seen in the screenshot and press **OK**:

Press **Next** to proceed:

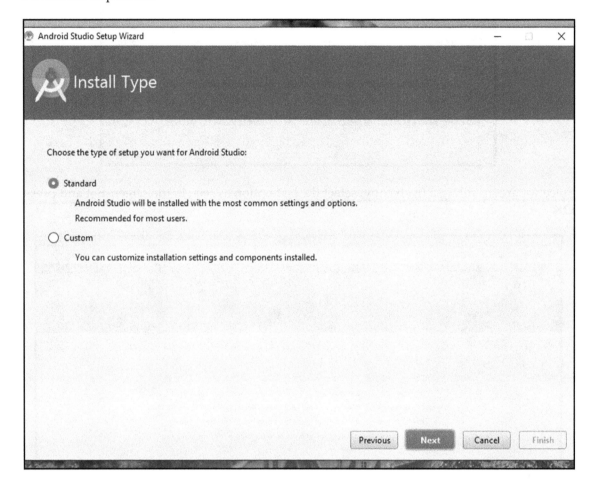

Select **Standard** installation for recommended settings:

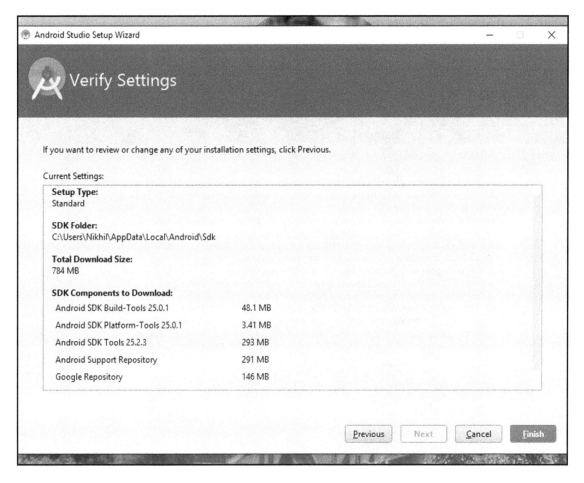

Press **Finish** to start downloading the required components.

Once you press **Finish**, your computer will start downloading the required components for Android N SDK, so ensure that your Internet connection is working and sit back and enjoy a cup of coffee while SDK gets installed on your system:

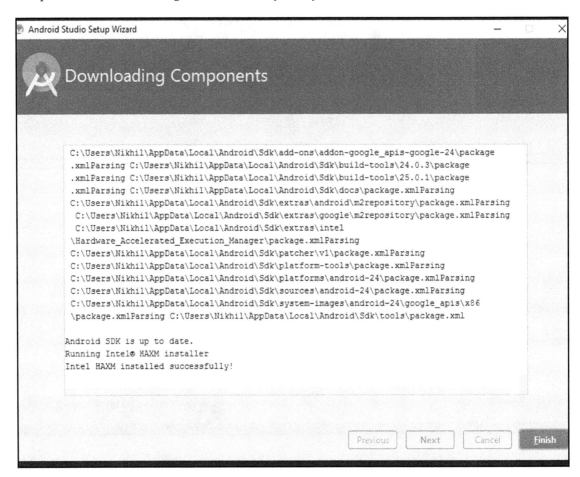

You have now successfully installed Android Studio and the components needed for Android N.

Once you are done downloading all the components of SDK, you will be ready to start Android Studio and will get the following screen menu:

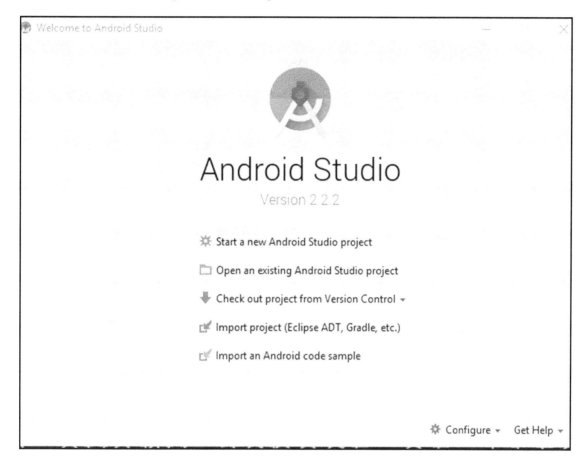

Congratulations! You are ready to start Android Studio now!

The nuts and bolts of Android

Before you start a new project in Android Studio, there are some basic concepts you must be familiar with. So, let's take a look at a few common terms we will be dealing with in our chapters.

Package names

The first thing that you will come across is something called a **Package Name**. It's quite easy to understand actually. A package name is simply like a reverse URL. Think of a package name as your app's domain name-just like a website, only in reverse. For instance, you can think of `www.google.com` as a website's domain name; in exactly the same way, the naming convention of an Android app is the reverse of a website. So, you can name your app something like `com.google.www`. There is no strict rule that says that your package name must start from `com`, but it is the most commonly accepted convention. You can also name your package name randomly using your own set of conventions, such as `abc.xyz.lmn`, `mygame.mycompany.myname`, and so on. Also, it is extremely important to note that package names must be unique and should not match the package name of any other existing app on the Google Play Store.

It is very important to choose a unique package name since the URL gets indexed by Google and is crucial for your game or app to be noticed on the Google Play Store. So, ensure that you use a unique package name for your game. Also, another interesting fact is that you can predict your app's URL even before it goes live if you have finalized it on your package name. For this reason, you cannot use the same package name of another app since it's already live on Google Play Store. Your app will be live according to the following URL convention:

```
https://play.google.com/store/apps/details?id=*package_name_here*
```

So, if your package name is `abc.xyz.lmn`, then your app's URL will be as follows:

```
https://play.google.com/store/apps/details?id=abc.xyz.lmn
```

Layouts

The next concept is Layouts. We will be dealing with Layouts in the next chapter but, just to give you a small introduction, let's provide a few examples. We will make a game, and in a game, we do not need to display the **status bar**, which means that we need to have a **Fullscreen Layout**. If you were making an app, then you probably would not mind allowing the status bar to be displayed on top. So, in this case, you can use a **Relative Layout** or **Linear Layout**. The really interesting aspect of this book is that, by the end, you will also have a basic idea of how to create a non-gaming app as well. So, it is highly recommended that you grasp the knowledge of the first three chapters properly.

Android Manifest file

Another important concept while making an Android app or a game is the `Android Manifest` file. To explain this file simply, it contains all the rules or, in more generalized terms, **Permissions** needed for an app. You must have observed on Google Play Store that, before you download any app, you are prompted with a dialog box that tells you what permissions are needed for the app to run. These permissions are basic rules that are needed to be fed in the `Android Manifest` file in order to have a transparency to let your user know what information is going to be taken from them. So, for instance, if an app requires access to the Internet, then it is required for the developer to make sure that they include the Internet permission in the manifest file. If the developer does not write this permission in the manifest file, then the app will not be able to access the said functionality, and the same goes for accessing contacts, gallery, camera, and everything else.

These are the three most important things you need to keep in mind before you start developing any Android game or an app.

In the next chapter, we will get started with our first project on Android Studio.

Summary

In this chapter, we have learned some basic information about Android as well as how to install Android Studio, which will help us on our journey to developing our apps. We also configured Android studio with components of Android N; you are now ready to get started with game development in Android.

Now that we have installed Android Studio, we will be learning how to execute/run our first program in the next chapter. Fasten your seatbelts, you are in for a ride!

2

Getting Familiar with Android Studio

This chapter will guide you through Android Studio and, by the end of this chapter, you will have successfully executed your first Android project. This will be an important chapter for your core understanding of the project structure of Android and will help you with the chapters that follow. You will learn the following:

- General project structure of an Android Project
- Default class explanation
- XML files and different types of XML files
- Setting up Android Emulator
- Executing your first Hello World program

Understanding the Android project structure

So, in our last chapter, we successfully set up our Android Studio with the required components to start working on our project right away. Let's get started. In order to learn about the Android project structure, we must first open up a new project.

Creating your first Android Studio project

Open your Android Studio, and click on **Start a new Android Studio project**, as shown in the following screenshot:

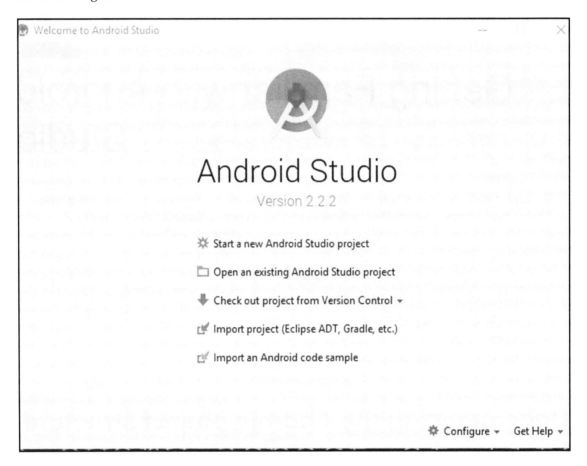

Once you start a new project, you will see the following screenshot:

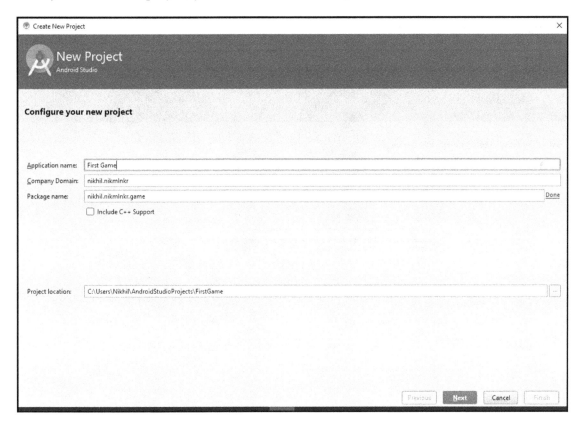

In this screen, fill in the details of your first Android app:

- **Application name** is the name of your app, which will appear on the icon of your phone when it is installed.
- **Company Domain** is an identifier for your app. Make sure that you keep this common throughout your apps for better organization and convention.
- **Package name** is another important unique identifier for your app. We learned about this in our `first chapter` and saw its naming conventions. Refer to that part if you have any doubts regarding package names and their naming.
- **Include C++ support** is optional. For the purpose of this book, we will keep it unchecked for now.
- **Project location** is the path where your project folder will be situated on your computer.

Once you are ready after filling in all these details, press **Next**.

Now, you will see this screen:

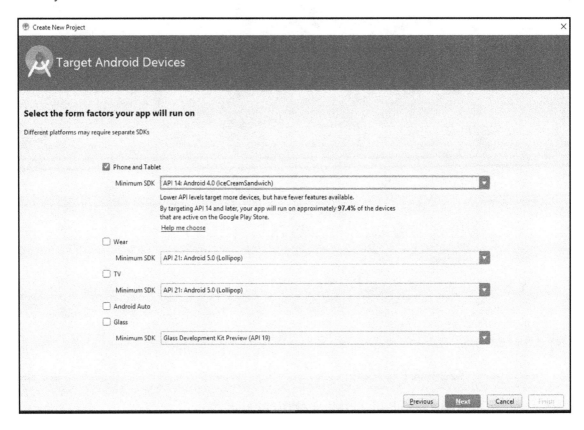

Different platforms for your Android app

For the purpose of this book, we will only work with **Phone and Tablet**. However, you can try experimenting with other platforms as well, once you are comfortable with the development cycle.

Minimum SDK is the OS version that would be required as a minimum factor to run your app. It is recommended that you select the lowest version for your app to run on as many OS versions as possible. However, do note that some functions are deprecated from further versions, so it is recommended that you use Minimum **API 14: Android 4.0 (IceCreamSandwich)** for hassle-free development. You will also be prompted about how many devices will be supporting your app, as you can see in the image. These figures are real time and change as the market share of an OS changes.

 If you want to know about the updated market share and some more interesting stuff related to various versions and their respective market share, you can visit the official Android website at `https://developer.android.com/about/dashboards/index.html`.

Once we select our **Fullscreen Activity**, press **Next**, as follows:

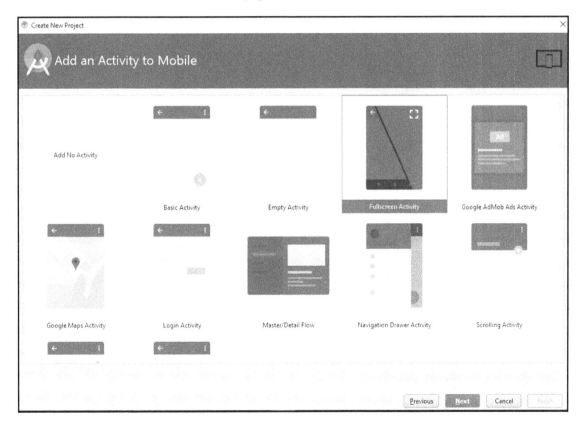

Different types of activities

You could say that an activity is simply a default layout. Since we are creating a game, we will use a **Fullscreen Activity**. As you can see in the preceding screenshot, you have many different activity options to choose from, which are quite self-explanatory on their own. So, let's select **Fullscreen Activity** and press **Next**:

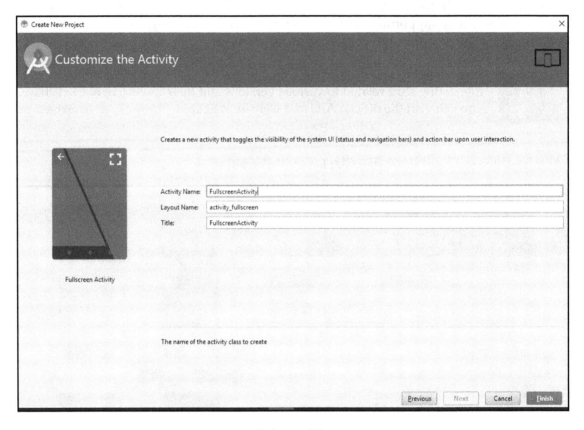

Naming your activities

Activity Name is the name of your Java class that will be generated. We will be working with this file a lot, so make sure that you remember it.

Layout Name is the name of the XML file, which will deal with how and what components will appear on your app visually. There are some naming conventions that need to be followed for XML files. They are as follows:

- Names must be all lowercase
- An underscore is used to separate two words instead of spaces.

- Always prefix the filename with the type of resources; for instance, if your XML file corresponds to an activity such as `MainActivity.java`, then your filename would be `activity_main.xml`.

- If you have a subitem of a specific group, such as a list item of main activity, then it can be named by suffixing it as `activity_main_list_item.xml`.

- This way, you can either suffix or prefix a keyword for your filename, as you prefer.

Title is the name that will appear on the top bar of your app. The title you use for the Activity and the class name of the main Activity might be different; titles may also contain spaces.

Once you are finished with all of this, click on **Finish** and give yourself a pat on the back. You have successfully learned how to create an Android Studio Project. After this, wait for a few seconds/minutes depending on your system's performance for your project to set up; once it is ready, you will see the following screen:

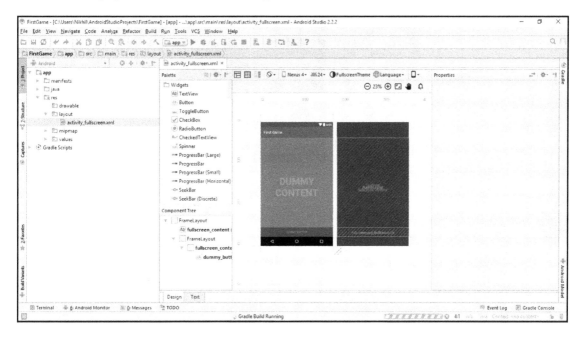

Our Android Studio Project Folder

If for some reason, you are not able to see this screen after waiting, then click on the **1: Project** option, which is vertically aligned below **FirstGame** on the upper-left corner of the screen.

Congratulations! You have successfully created your first app now. At this point, you can go ahead and run the project and see how it runs on the emulator; however, before we do that, let's walk through the project folder structure of this project. Let's expand each folder to understand it further. Now, this part is very important, since this will serve as a foundation for almost every basic thing you will be doing in Android Studio, so make sure that you understand this properly.

Project Structure of an Android Project

Click on the small arrow on the left-hand side of each folder to expand its view:

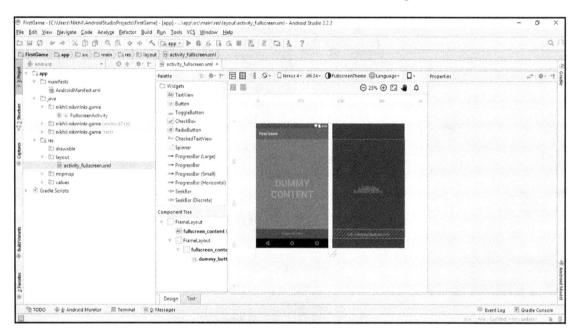

Various folders inside the project folder

As you can see, we have multiple folders in our project folder. Our main project folder contains three subfolders, which are `manifest`, `java`, and `res`; we will explore each of these folders individually now:

- `manifest`: This folder contains your `AndroidManifest.xml` file, which is responsible for giving permissions to your app, as we learned in the preceding chapter.

- `java`: This folder contains all your `.java` files, which are your Java code files. Generally, this folder has multiple subfolders that have your package name at the top, and within them, are source code files. You can observe in the preceding screenshot that we have our `java` folder, and we have a folder named `nikhil.nikmlnkr.game` within that, which is actually the package name of our game, and we have our `FullScreenActivity` within that.

- `res`: This folder contains all your resources. `res` here simply means resources that can include anything from simple string data to images to complex XML layouts. Simple terms, most of the things you see on the screen are stored here. You can design your front-end with the help of resources that you keep in the `res` folder.

Now that we know what these folders are, let's start designing our first program. Also, to stay true to our programming background, let's start with the famous `Hello World` example. Things are going to get really interesting from this point on. By now, you have grasped pretty much all the basics you need to start development. Now, in a matter of a few minutes, you will be ready with your very first Android app that you can run on your phone. So, without further ado, let's begin!

If you can see a screen with two big blue windows saying **DUMMY CONTENT,** as seen in the preceding screenshot, you are good to start; otherwise simply navigate to the `res/layout/` folder and double-click on `activity_fullscreen.xml` to open the XML layout in which we will be working now.

Now, you can see many options in front of you; don't get overwhelmed or scared. You will soon learn and master the art of creating amazing user interfaces with the tools in front of you.

Creating our Hello World! program

So, let's create our first `Hello World` program. As you can see, you have a **Palette** in front of you besides the blue **DUMMY CONTENT** window. Simply drag and drop the **TextView** component onto your **DUMMY CONTENT** screen, as you can see in the following screenshot:

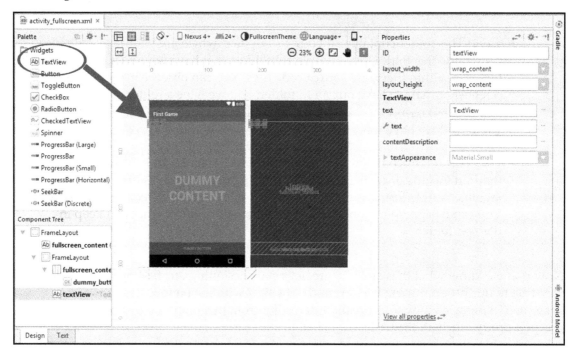

Dragging and dropping **TextView** component onto our screen

Notice the new **Properties** window when you successfully place **TextView** in the **DUMMY CONTENT** screen.

Now, you can see a blank **Text View** component on your screen, but there's nothing written on it. To have something written on it, we need to modify its properties a little. Check out the **Properties** on the right-hand side of the screen and, in the **text** component over there, type in `Hello World!`. Also, you can see that our text is very small. Let's change that to make it a little bigger so that it can be clearly visible to us. Locate the last option in the **Properties** window named **textAppearance**. Click on the drop-down menu next to it, and select any option that suits your choice from there. For this example, we will use **AppCompat.Headline**; however, you can choose whatever you want to. Once you are done with it, you will see something like this:

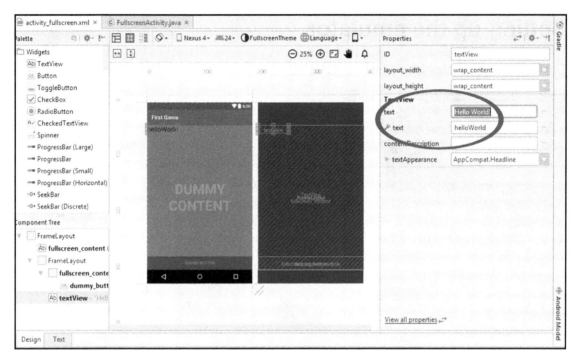

Your first Hello World! program is ready!

That's it! You are now ready to execute your program! Easy peasy, lemon squeezy, right? Let's move on to actually executing this code now. For this purpose, we will need to run something known as the **Emulator**. An emulator is simply a virtual device that will act as an Android device on your PC, so you don't have to actually test your app on an Android device every time. You will need to follow a few steps to set up an emulator, so let's get started with it. You will only have to set up the emulator once, and next time, it will be ready for you.

Setting up the emulator

Click on the green play icon to start executing your first program and set up your emulator:

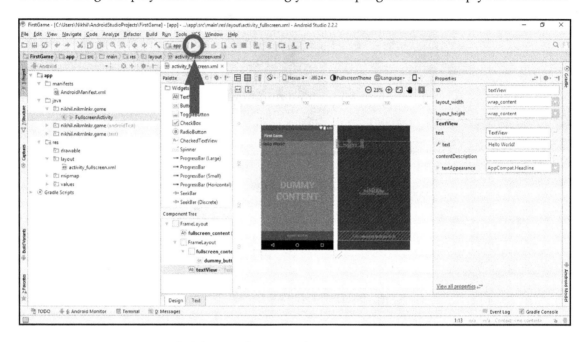

The play button can be found on the top of the screen as marked in this image

Now, click on **Create New Virtual Device** and make sure that **Use same selection for future launches** is checked:

Select a device of your choice, and click on **Next**:

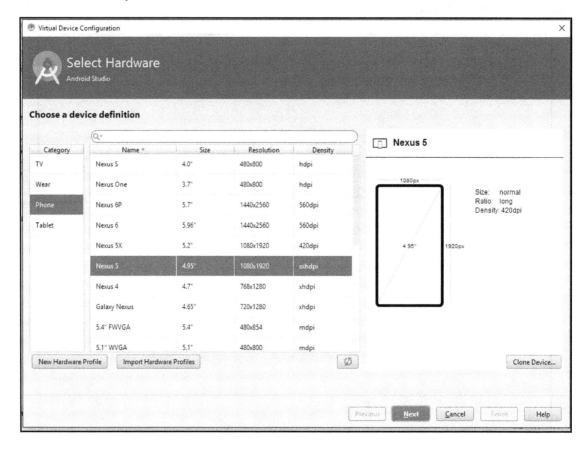

Select **Nougat** with **API Level 24**, and click on **Next**:

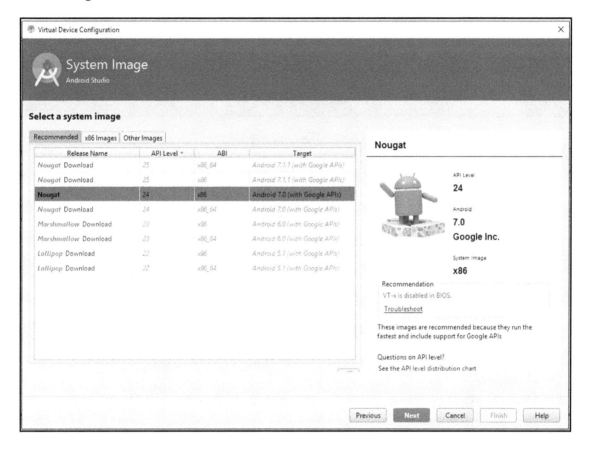

Give a name to your emulator (optional) in the **AVD Name** field and select a **Portrait -
Startup orientation** and click on **Finish**:

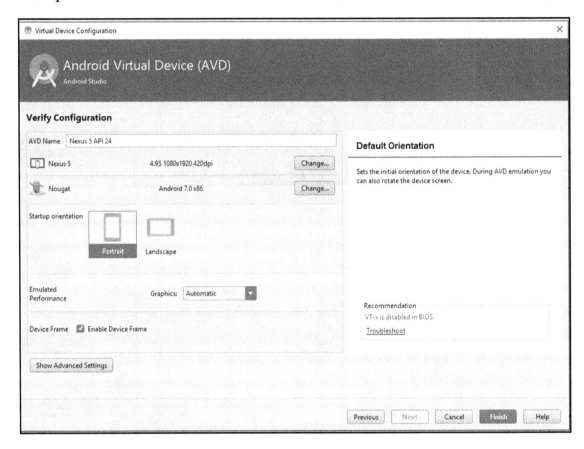

You can see that the latest device that you created will now appear in the list of **Available Virtual Devices**; select it and click on **OK**:

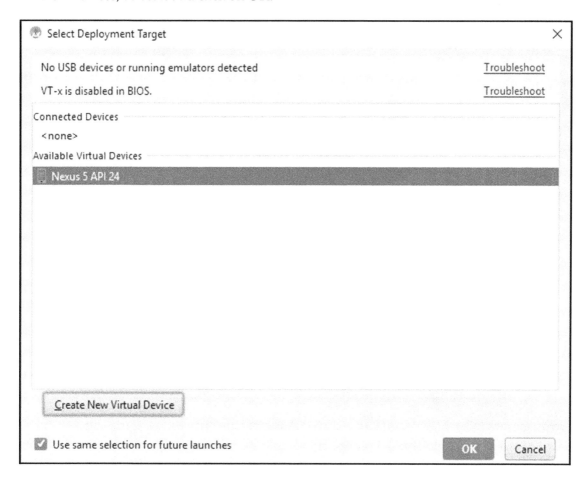

Click on **OK** after selecting your device

ONLY FOR WINDOWS SYSTEMS

Now you may get the following message, as seen in the screenshot:

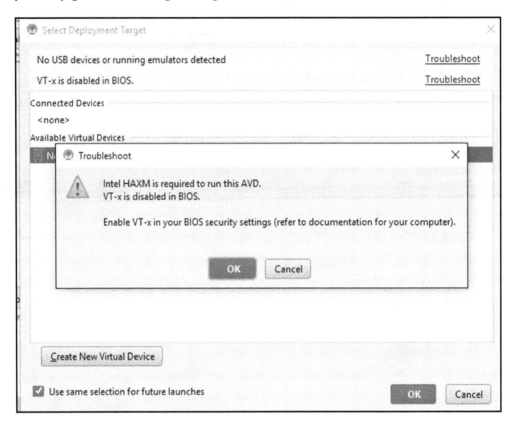

If you get an **ONLY FOR WINDOWS SYSTEMS** message, then follow these steps; otherwise, you can skip this part:

1. Enter BIOS by restarting your computer and pressing *Delete*, *Esc*, or *F1* depending on your system.
2. Go to **Processor/Chipset** settings.
3. Enable **Virtualization Technology.**

This can happen due to a couple of problems. HAXM problems often occur because of Intel chipsets. The **Intel Hardware Accelerated Execution Manager** (**Intel® HAXM**) is a hardware-assisted virtualization engine (hypervisor) that uses **Intel Virtualization Technology** (**Intel® VT**) to speed up Android app emulation on a host machine. In combination with Android x86 emulator images provided by Intel and the official Android SDK Manager, HAXM allows for faster Android emulation on Intel VT-enabled systems:

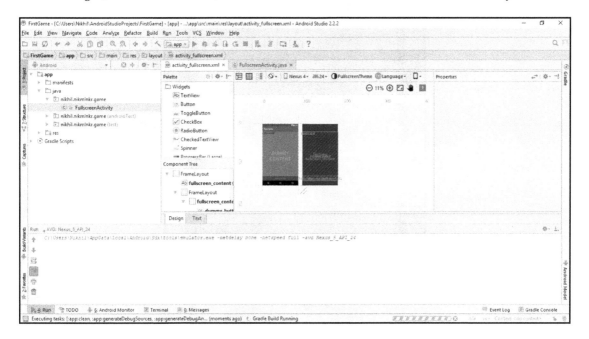

Once you do this, a small window will open below the screen, and your build process will start. Wait for some time and your emulator will open up:

Congratulations! Your Hello World! program is running successfully.

Give yourself a pat on the back. You have now successfully executed your first program.

Summary

In this chapter, you have gone through our Android Project Structure as well as developing a basic understanding of how to create elements using the XML file. We also learned how to set up the Android Emulator and executed our Hello World! program.

In the next chapter, we will take a look at how to manage inputs and we will dive deeper into understanding XML files, which will serve as a core foundation for making our games. We will also learn how to link your XML files to your source code, change texts, and other cool stuff.

3
Managing Inputs

Now that we have learned how to run a program on the Android Emulator, it is time for us to do some more cool stuff, which will equip us with the knowledge required to make our game. In this chapter, we will continue with XML files and will step into the territory of taking user inputs. In a gist, we will be learning the following:

- Exploring different types of XML files and the resource folder further
- Creating buttons and linking them to take inputs
- Working with accelerometer readings
- Mobile touch inputs

So, let's get started.

Resource folder in detail

In the last chapter, we used the `activity_fullscreen.xml` file to edit the frontend of our application. Now, we will take a look at some more of these XML files and understand how they can be useful to us for making games. To understand these type of files, we must first know a little bit about them. To start off, the very basic information about XML is that it's a short form of *Extensible Markup Language*. Now, if you have studied HTML, you will know that its full form is quite similar to it--HyperText Markup Language. It's quite similar in syntax as well, but the function of an XML file is to hold data. If you go by the definition of an XML file, it goes something like this: XML is a software and hardware-independent tool for storing and transporting data.

You can read more about XML files at `https://en.wikipedia.org/wiki/XML`.
We have not yet seen an XML file in the code yet, so let's do that. Click on **Text** as shown in the following screenshot:

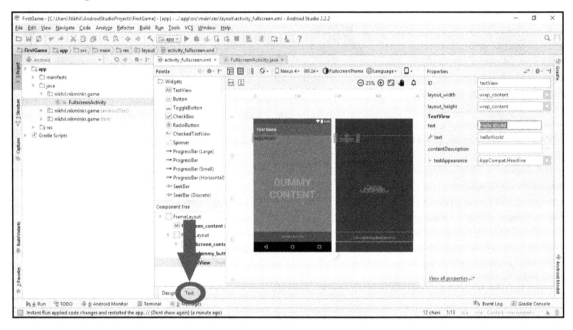

The **Text** mode for code editing is just besides the **Design** button

Now, you can actually see the XML code opened up:

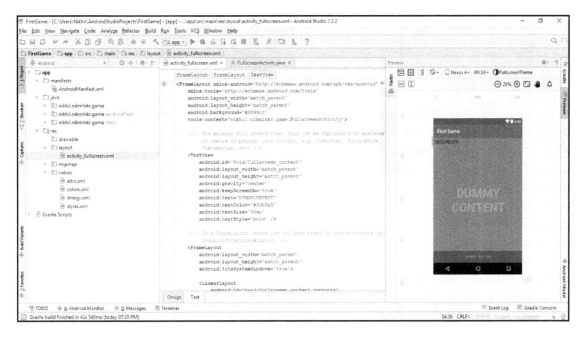

This is your default XML code

Pay close attention to this code, and you will find something like this:

```
<TextView
    android:text="Hello World!"
    android:layout_width="wrap_content"
    android:layout_height="wrap_content"
    android:id="@+id/TextView"
    tools:text="helloWorld"
    android:textAppearance="@style/TextAppearance.AppCompat.Headline" />
```

If you see this, you can observe your `Hello World!` text on the very first line of this code. The data that we changed visually in our previous chapter can be changed through code over here. It is almost the same for all components, and you will learn about different components as you practice further.

Now, this is not the only type of XML file. As we read in the definition, XML files are used to store data. Let's see the other kinds of XML files, which can be used to store data. We can use these files for storing game scores, filenames, text data, and so much more. Let's take one such new type of XML file, which is already available in our project folder to understand further:

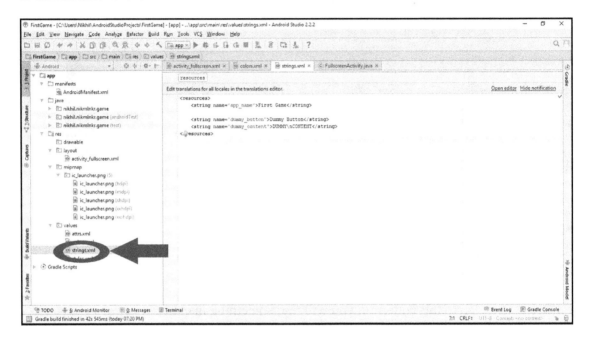

The strings.xml file contains all your string data and is in the **res/values/** folder

Navigate to the `app/res/values/` folder and double-click on the **strings.xml** file.

You can see the code for this file when you open it. Here, you can observe that there are multiple values and an ID to each value in the form of a `name`. Observe carefully the second line that reads as follows:

```
<string name="app_name">First Game</string>
```

Remember this name? We had set this as our app name when we started the project. It is stored in the value `app_name`. Also, if you go back and search for `app_name` using *Ctrl + F* on your `activity_fullscreen.xml` file, you will find this entry. Go ahead and explore a little for yourself.

Also, check out the other files to get an understanding. Here's a basic explanation for the four XML files in our project folder:

- `attrs.xml`: This declares custom theme attributes that allow changing the styles that are used for button bars, depending on the API level
- `colors.xml`: This defines colors that can be used in hex code
- `strings.xml`: This holds data for all string-related values
- `styles.xml`: This sets the base theme of the application

So, that's it about XML files. Let's now move on to something even more interesting--inputs.

Taking user inputs

There are multiple ways in which you can take inputs from your Android device. Here are a few ways:

- UI buttons: Buttons drawn on your app's UI
- Hardware buttons: The keys on your Android device
- Touch screen inputs: Touches mapped based on the screen coordinates
- Accelerometer readings: Motion sensor readings

We will be taking a look at each of these input types. So, let's start with the very first type of input, UI buttons.

Button input

The button input is one of the most common type of components used in an Android project. Let's come back to our design mode and create a button on the screen:

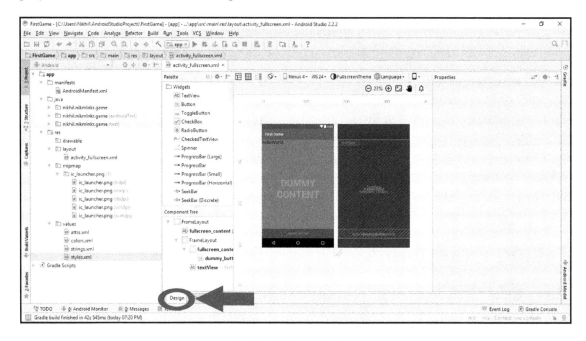

Reverting back to **Design** mode from **Text** mode

Click on the **Design** button and switch back to our visual editor mode on our **activity_fullscreen.xml** file.

Here, we will need to do some changes to our XML file. Follow the steps as mentioned:

1. From the **Palette**, scroll down and find **Linear Layout(Horizontal)** and drag and drop it inside the **fullscreen_content_controls.**
2. In the **Component Tree** window below the **Palette**, drag and drop your **TextView** under your newly created **LinearLayout(horizontal)**.
3. Select the **Button** component from your **Palette** and drag and drop it in into your **Component Tree** window's **LinearLayout(horizontal)**.

Once you do this, you will get an output like this:

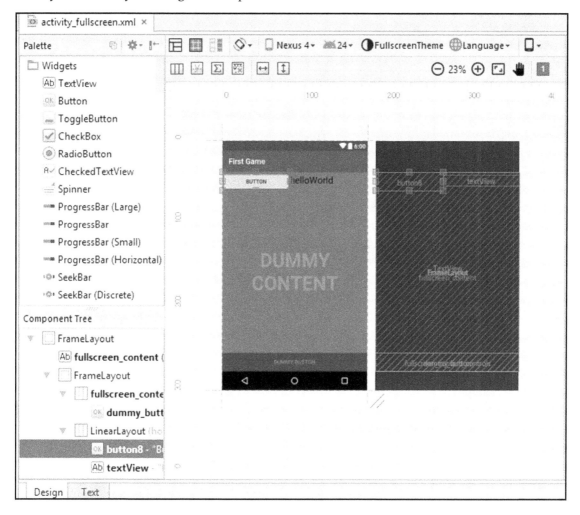

You now have a button and text on the screen.

If we had not used the **Linear Layout** component, then our button and text would have overlapped each other since if you observe in your **Component Tree** our **TextView** was initially inside **FrameLayout**, which does not have alignment options. You can try dragging your button and text inside the **FrameLayout** and see for yourself.

Now, let's proceed toward making this button work. Click on your newly created button, and in the properties window on right side, take a look at the **ID** property. This is the property that lets your code know which button to interact with; change it to myFirstButton:

You have now given a custom ID to your button.

Repeat the same for your **TextView** by changing its ID to `myTextView`. Keep these two IDs in mind, and let's move on to our next step. Now, we will actually link this visual button to our code, so we can change the text of our **TextView** component. Go to your `app/java/nikhil.nikmlnkr.game` folder and open up the **FullscreenActivity**.

Note that `nikhil.nikmlnkr.game` is the package name used in this book. Your package name might be different based on what you set it initially, so navigate to the folder based on your package name. For the purpose of this book, we will assume the package name is `nikhil.nikmlnkr.game`.

After we open that file, we will have opened the main Java code for our application. You can see that there is already a lot of code written over there. Don't get overwhelmed by it because soon you will understand on your own what each code block is about. Let's start by writing some code of our own to get started to link our button to our code file; search for the `void onCreate` function.

The `onCreate()` method is where you initialize your activity. If the activity is started and the application is not loaded, then both `onCreate()` methods will be called. You can initialize your variables and methods over here.
You can see something like this:

```
@Override
protected void onCreate(Bundle savedInstanceState) {
    super.onCreate(savedInstanceState);

    setContentView(R.layout.activity_fullscreen);

    mVisible = true;
    mControlsView = findViewById
    (R.id.fullscreen_content_controls);
    mContentView = findViewById(R.id.fullscreen_content);

    // Manually show or hide the System UI
    mContentView.setOnClickListener(new View.OnClickListener(){
        @Override
        public void onClick(View view) {
            toggle();
        }
    });

    /* Upon interacting with UI controls, delay any scheduled hide()
    operations to prevent the jarring behavior of controls going away
while
```

```
        interacting with the UI.*/

        findViewById(R.id.dummy_button).
        setOnTouchListener(mDelayHideTouchListener);
    }
```

Now, modify it to look somewhat like this; changes are marked in bold:

```
@Override
protected void onCreate(Bundle savedInstanceState) {
    super.onCreate(savedInstanceState);

    setContentView(R.layout.activity_fullscreen);
    // Declare variable references for our TextView and Button with
    // their IDs
    final TextView tv =
    (TextView)findViewById(R.id.myTextView);
    final Button button =
    (Button) findViewById(R.id.myFirstButton);

    // Make an OnClickListener to listen to button click events
    button.setOnClickListener(new View.OnClickListener() {
        public void onClick(View v) {
            // Perform action on click
            tv.setText("Button Clicked");
        }
    });
    mVisible = true;
    mControlsView =
    findViewById(R.id.fullscreen_content_controls);
    mContentView =
    findViewById(R.id.fullscreen_content);


    // Manually show or hide the System UI
    mContentView.setOnClickListener(new View.OnClickListener(){
        @Override
        public void onClick(View view) {
            toggle();
        }
    });

    // Upon interacting with UI controls, delay any scheduled hide()
    // operations to prevent the jarring behavior of controls going
    // away while interacting with the UI.
```

```
    findViewById(R.id.dummy_button).
    setOnTouchListener(mDelayHideTouchListener);
}
```

By writing this code, you have done the following:

- Linked your button's ID to your code
- Linked your text view's ID to your code
- Created a button click listener, which is needed, to take input from a button

Once you are ready with this, run your app on the emulator with the steps we learned in the preceding chapter:

On pressing our **Button**, the text will change to **Button Clicked**, as we coded in our file.

Hardware button input

Now, let's test out our hardware buttons along with a Toast example. Don't feel hungry yet reading Toast! It's not the one that you are going to eat. A Toast is a message that gets displayed on your screen for a few seconds and disappears. You will see what it is in a few minutes. So, now open up your `FullscreenActivity.java` file where you typed your `onCreate` code, and after your `onCreate` method, type the following:

```
@Override
public void onBackPressed() {
    // your code.
    Toast.makeText(FullscreenActivity.this,
    "Back button pressed", Toast.LENGTH_SHORT).show();
}
```

Observe the code here. In the first line, we have used the keyword `@Override`. This is because we are overriding the parent class functionality, which is the default Android behavior to do something else that we want. If you do not override the function, then by default Android will close the app since that is the function of the back button.

If you are getting an error on Toast with a red underline, then click on the text where you are getting the error and then press *Alt + Enter*. This will add the missing imports that are required to be included for running it. Once you are done with it, run your app again on the emulator, and you will see the following output:

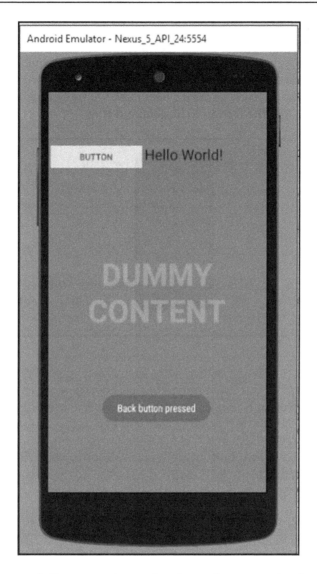

So, now you have successfully mapped your hardware button to display a Toast message. Bravo! Let's move ahead and try to track our touch coordinates.

Touch input

Now that we have seen how to map our hardware buttons, let's dive deeper into the most commonly used input on Android: touch input. However, before we understand touch input, we must understand the coordinate system used to track touches on the screen. Let's take a look at the following illustration to understand this:

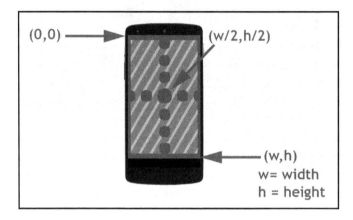

Coordinates system in Android

In order to track our touches, we must have a basic understanding of how screen coordinates are mapped on our device. As you can see in the preceding image, our screen's coordinates start from (0,0) at the top-left corner and end at the bottom-right corner with **(w,h)** where w is the screen width and h is the screen height. So, let's say if your phone's resolution is 480 x 850, then your bottom-right corner coordinates will be w=480, h=850. Thus, your extreme coordinates will be mapped as (480,850). Just keep this in mind, and you will understand the reason why we need to know this shortly.

Now, open up your XML file and drag and drop another **TextView** component:

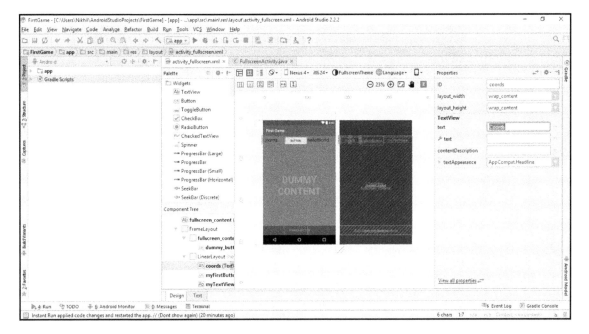

Give it an **ID** coords and **text** as Coords:

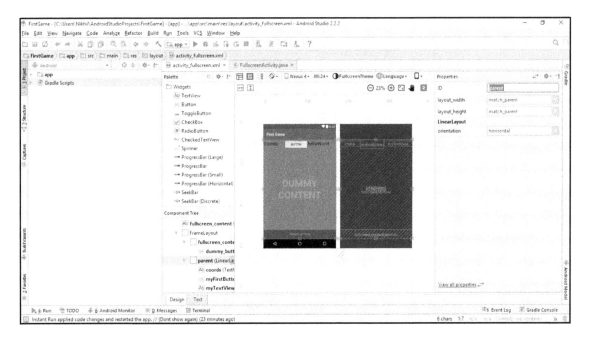

Now, click on your **LinearLayout** from the **Component Tree** view and give it an **ID** as
parent.

Now, go to your `FullscreenActivity.java` **code file and modify your** `onCreate`
function to look like this:

```
@Override
protected void onCreate(Bundle savedInstanceState) {
    super.onCreate(savedInstanceState);

    setContentView(R.layout.activity_fullscreen);
    final TextView tv =
    (TextView) findViewById(R.id.myTextView);
    final Button button =
    (Button) findViewById(R.id.myFirstButton);

    button.setOnClickListener(new View.OnClickListener() {
        public void onClick(View v) {
            // Perform action on click
            tv.setText("Button Clicked");
        }
    });
    mVisible = true;
    mControlsView =
    findViewById(R.id.fullscreen_content_controls);
    mContentView =
    findViewById(R.id.fullscreen_content);
    final LinearLayout parent =
    (LinearLayout) findViewById(R.id.parent);

    final TextView text = (TextView) findViewById(R.id.coords);

    parent.setOnTouchListener(new View.OnTouchListener() {
        public boolean onTouch(View v, MotionEvent ev) {
            text.setText
            ("Touch at " + ev.getX() + ", " + ev.getY());
            return true;
        }
    });

    // Set up the user interaction to manually show or hide the system UI.
    mContentView.setOnClickListener(new View.OnClickListener() {
        @Override
        public void onClick(View view) {
            toggle();
        }
    });
```

```
// Upon interacting with UI controls, delay any scheduled hide()
// operations to prevent the jarring behavior of controls going
// away while interacting with the UI.

findViewById(R.id.dummy_button).
setOnTouchListener(mDelayHideTouchListener);
}
```

To give an explanation for the preceding code, what we have done over here is quite simple: we have declared a reference for our **LinearLayout** as `parent` and then another reference for our newly created **coords(TextView)**. Now, after our declaration, we instructed the `parent` class to have a `Touch listener` property on it. This will help us get the coordinates that we touched. As you can see in the preceding code, we have a `MotionEvent` variable `ev`, which will give us the coordinates. Then, in the next line, we fetched the value in the form of x and y coordinates, which will then be set as text on our **coords(TextView)**.

Go ahead and run your code now, and when the emulator starts, try clicking randomly anywhere; that will give you coordinates that you just touched:

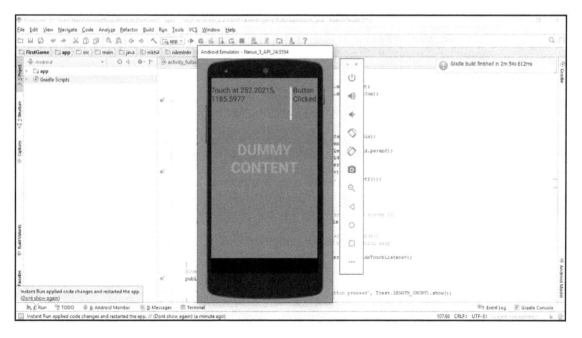

It will show you where you touched on the screen.

Now, by this time, you must have observed that it is pretty tedious to drag and drop and align all the texts properly, right? That is because we are working with **LinearLayout**. There is a way through which you can organize all your texts properly. That is through **RelativeLayout**. Let's take a look at a working example to help you get a better understanding of the same. Go to the **TextView** of your XML file as we learned to do in the preceding chapter, and wherever you see **LinearLayout** change it to **RelativeLayout**. Here is a reference code for the same:

```xml
<RelativeLayout
    android:orientation="horizontal"
    android:layout_width="match_parent"
    android:layout_height="match_parent"
    android:id="@+id/parent">

    <Button
        android:text="Button"
        android:layout_width="wrap_content"
        android:layout_height="wrap_content"
        android:id="@+id/myFirstButton"
        android:layout_weight="1" />

    <TextView
        android:text="Hello World!"
        android:layout_width="wrap_content"
        android:layout_height="wrap_content"
        android:id="@+id/myTextView"
        tools:text="helloWorld"
        android:textAppearance="@style/TextAppearance.AppCompat.Headline"
        android:layout_weight="1"
        android:layout_alignBaseline="@+id/myFirstButton"
        android:layout_alignBottom="@+id/myFirstButton"
        android:layout_toRightOf="@+id/myFirstButton"
        android:layout_toEndOf="@+id/myFirstButton" />

    <TextView
        android:text="Coords"
        android:layout_width="wrap_content"
        android:layout_height="wrap_content"
        android:id="@+id/coords"
        android:layout_weight="1"
        android:textAppearance="@style/TextAppearance.AppCompat.Headline"
        android:layout_below="@+id/myFirstButton"
        android:layout_alignRight="@+id/myFirstButton"
        android:layout_alignEnd="@+id/myFirstButton" />
</RelativeLayout>
```

Now, you have successfully converted your **LinearLayout** to **RelativeLayout**, which will give you more control over your design options. You can now adjust your components on screen with more ease, and so you can align them nicely as follows:

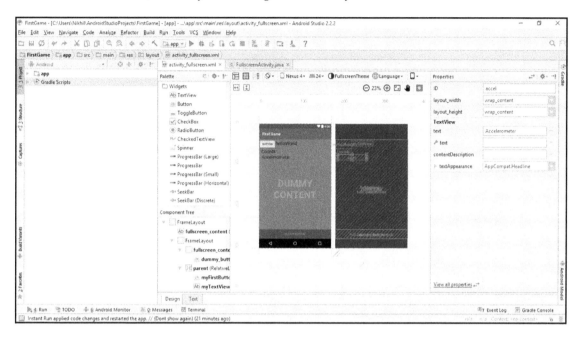

Once you do this, you can easily adjust your texts by simply dragging them anywhere you want. You won't mess up the layouts when you use **RelativeLayouts**. Now, let's move on to the next example for getting accelerometer inputs, and take another **TextView** component on screen for that as you see in the preceding code and give it an **ID** `accel` with a **text** `Accelerometer`. However, before we proceed with the code for the same, there is a minor change that we need to do in our existing code.

Since we just changed our **LinearLayout** to **RelativeLayout**, we also have to reference the same for our touch function in our Java code file. Open up your Java code file, and there you will observe that it is already giving you an error prompt on **LinearLayout**. That is because there is actually no **LinearLayout** in our project with the ID **parent** because we changed it to **RelativeLayout**. Don't worry about this though; simply type in **RelativeLayout** instead of **LinearLayout**, and you are sorted. Here is how:

```
@Override
protected void onCreate(Bundle savedInstanceState) {
    super.onCreate(savedInstanceState);

    setContentView(R.layout.activity_fullscreen);
```

```
final TextView tv =
(TextView) findViewById(R.id.myTextView);
final Button button =
(Button) findViewById(R.id.myFirstButton);

button.setOnClickListener(new View.OnClickListener() {
    public void onClick(View v) {
        // Perform action on click
        tv.setText("Button Clicked");
    }
});
mVisible = true;
mControlsView =
findViewById(R.id.fullscreen_content_controls);

mContentView =
findViewById(R.id.fullscreen_content);

final RelativeLayout parent =
(RelativeLayout) findViewById(R.id.parent);

final TextView text =
(TextView) findViewById(R.id.coords);
parent.setOnTouchListener(new View.OnTouchListener() {
    public boolean onTouch(View v, MotionEvent ev) {
        text.setText
        ("Touch at " + ev.getX() + ", " + ev.getY());
        return true;
    }
});
// Set up the user interaction to manually show or hide the system
UI.
mContentView.setOnClickListener(new View.OnClickListener(){
    @Override
    public void onClick(View view) {
        toggle();
    }
});

// Upon interacting with UI controls, delay any scheduled hide()
// operations to prevent the jarring behavior of controls going
// away while interacting with the UI.
findViewById(R.id.dummy_button).
setOnTouchListener(mDelayHideTouchListener);
}
```

Run your app once, and you can see how neat it looks now:

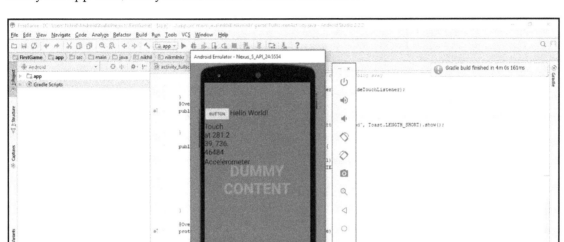

Now, we have successfully made pretty neat adjustments to our layout as well as implemented a touch listener to detect touches.

Now, we move on to our final type of input for this chapter, *The Accelerometer input*.

The Accelerometer input

We will now take a look at the accelerometer component on Android. If you don't know what an accelerometer is, it is something that is used to detect movement in Android. In layman terms, we can call it the motion sensor. The best example for this is any racing game that lets you control the car based on your phone's movement. This is something really interesting, which you can use in further chapters to apply motion to objects, so make sure that you grasp this properly. Type in the code you see in the following block starting from `sensorManager`:

```
@Override
protected void onCreate(Bundle savedInstanceState) {
    super.onCreate(savedInstanceState);

    setContentView(R.layout.activity_fullscreen);
```

```
final TextView tv =
(TextView) findViewById(R.id.myTextView);

final Button button =
(Button) findViewById(R.id.myFirstButton);

button.setOnClickListener(new View.OnClickListener() {
    public void onClick(View v) {
        // Perform action on click
        tv.setText("Button Clicked");
    }
});
mVisible = true;
mControlsView =
findViewById(R.id.fullscreen_content_controls);
mContentView =
findViewById(R.id.fullscreen_content);

final RelativeLayout parent =
(RelativeLayout) findViewById(R.id.parent);
final TextView text =
(TextView) findViewById(R.id.coords);

parent.setOnTouchListener(new View.OnTouchListener() {
    public boolean onTouch(View v, MotionEvent ev) {
        text.setText
        ("Touch at " + ev.getX() + ", " + ev.getY());
        return true;
    }
});
SensorManager sensorManager =
(SensorManager) getSystemService(Context.SENSOR_SERVICE);
Sensor sensor =
sensorManager.getDefaultSensor(Sensor.TYPE_ACCELEROMETER);

sensorManager.registerListener(new SensorEventListener() {
    @Override
    public void onSensorChanged(SensorEvent event) {

        float x = event.values[0];
        float y = event.values[1];
        float z = event.values[2];
        TextView acc = (TextView) findViewById(R.id.accel);
        acc.setText("x: "+x+", y: "+y+", z: "+z);

    }

    @Override
```

```
        public void onAccuracyChanged(Sensor sensor,
        int accuracy){
        }
    }, sensor, SensorManager.SENSOR_DELAY_FASTEST);
    // Set up the user interaction to manually show or hide the system
UI.
    mContentView.setOnClickListener(new View.OnClickListener(){
        @Override
        public void onClick(View view) {
            toggle();
        }
    });
    // Upon interacting with UI controls, delay any scheduled hide()
    // operations to prevent the jarring behavior of controls going
    // away while interacting with the UI.

    findViewById(R.id.dummy_button).
    setOnTouchListener(mDelayHideTouchListener);
}
```

Again, over here if you get an error related to import statements while typing, then press *Alt + Enter* and you will be prompted with suggestions. Select the corresponding import statement from the list to solve the error. Now, the code will be explained. We made a variable for our `sensorManager` component of Android, which is responsible for tracking our `accelerometer` values. As you can see after the line following it, we have taken the sensor type as `TYPE_ACCELEROMETER`. Next, we created another listener for our component and then we took in the x, y, and z values. After that, we simply took a reference to the **accel TextView** component in our XML file and set its text to display those values.

Now, it is obvious that you won't be able to see accelerometer values changing on your emulator, so you will have to test it on your mobile device. However, just for curiosity's sake, try and run the code on your computer:

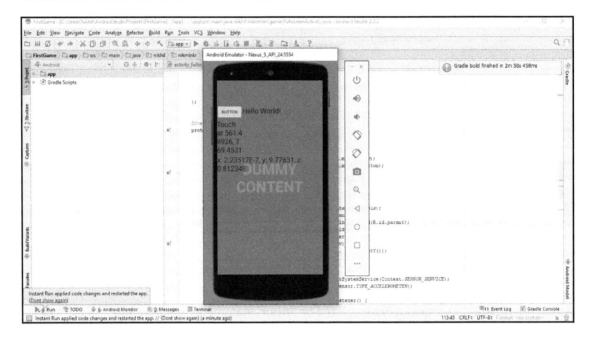

You will see some values, but they will not be changing because our PC does not have a motion sensor.

So, let's run our app on our mobile now. We can do this in two ways:

- Build and deploy directly to our device through USB, which will allow us to run the app directly on a physical device
- Build an apk and then transfer the apk file to our phone and install it

Let's start!

Building and deploying through USB

To make your device ready for debugging you need to enable your device for debugging first. In order to enable your device for debugging in **Developer options**. To access these settings, open the Developer options in the system settings. On Android 4.2 and higher versions, the Developer options screen is hidden by default. To make it visible, go to **Settings** | **About phone** and tap on **Build number** seven times:

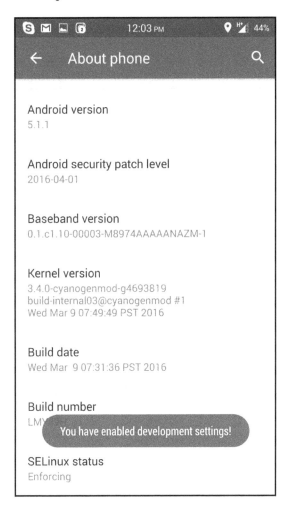

After this, return to the previous screen to find Developer options at the bottom:

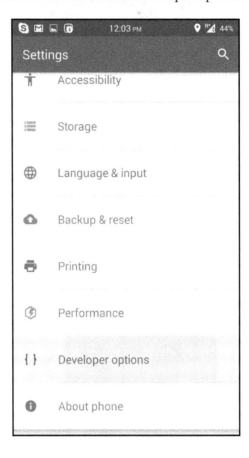

Now, enable the USB debugging option on your device:

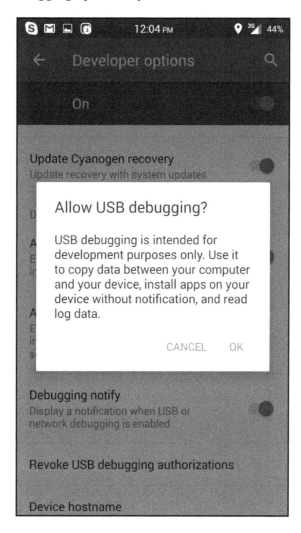

With an Android-powered device, you can develop and debug your Android applications just as you would on the emulator. Before you can start, there are just a few things to do:

Verify that your application is debuggable in your manifest or `build.gradle` file.

In the build file, make sure that the `debuggable` property in the `debug` build type is set to true. The build type property overrides the manifest setting:

```
android {
    buildTypes {
        debug {
            debuggable true
        }
```

In the `AndroidManifest.xml` file, add `android:debuggable="true"` to the `<application>` element.

 Note: If you manually enable debugging in the manifest file, ensure that you disable it in your release build (your published application should usually *not* be debuggable).

1. Enable **USB debugging** in the device system settings, under **Settings | Developer options**.

1. Note: On Android 4.2 and newer versions, **Developer options** is hidden by default. To make it available, go to **Settings | About phone** and tap on **Build number** seven times. Return to the previous screen to find **Developer options**.

2. Set up your system to detect your device:
   - If you're developing on Windows, you need to install a USB driver for ADB. For an installation guide and links to OEM drivers, refer to the `OEM USB Drivers` document.
   - If you're developing on macOS X, it just works, so skip this step.
   - If you're developing on Ubuntu Linux, you need to add a udev rules file that contains a USB configuration for each type of device you want to use for development. In the rules file, each device manufacturer is identified by a unique vendor ID, as specified by the ATTR{idVendor} property. For a list of vendor IDs, refer to `USB Vendor IDs`, as follows. To set up device detection on Ubuntu Linux, perform the following:
     1. Log in as root and create this file: `/etc/udev/rules.d/51-android.rules`.

1. Use the following format to add each vendor to the file:
   ```
   SUBSYSTEM=="usb", ATTR{idVendor}=="0bb4",
   MODE="0666", GROUP="plugdev"
   ```

   In this example, the vendor ID is for HTC. The MODE assignment specifies read/write permissions, and GROUP defines which Unix group owns the device node.

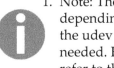

1. Note: The rule syntax may vary slightly depending on your environment. Consult the udev documentation for your system as needed. For an overview of rule syntax, refer to this guide at `writing udev rules`.

2. Now, execute `chmod a+r /etc/udev/rules.d/51-android.rules`.

Note: When you connect a device running Android 4.2.2 or higher versions to your computer, the system shows a dialog asking whether to accept an RSA key that allows debugging through this computer. This security mechanism protects user devices because it ensures that USB debugging and other ADB commands cannot be executed unless you're able to unlock the device and acknowledge the dialog. This requires that you have ADB version 1.0.31 (available with SDK Platform-tools r16.0.1 and higher) in order to debug on a device running Android 4.2.2 or higher versions.

When plugged in over USB, you can verify that your device is connected by executing ADB devices from your SDK platform-tools/ directory. If connected, you'll see the device name listed as a device.

On pressing run or debugging your application, you will be presented with a Device Chooser dialog that lists the available emulator(s) and connected device(s). Select the device on which you want to install and run the application.

If using the **Android Debug Bridge** (**ADB**), you can issue commands with the –d flag to target your connected device.
You can find the vendor IDs at `https://developer.android.com/studio/run/device.html`.

Building an apk and installing on device

From the top task bar where you see **File**, **Edit**, and so on, click on **Build** and then click on **Generate APK**. Once you have done this, it will start generating an apk for you. Once it is finished right-click on your **app** folder, which is on the left-hand side, as seen in the following screenshot, and click on **Show in Explorer**:

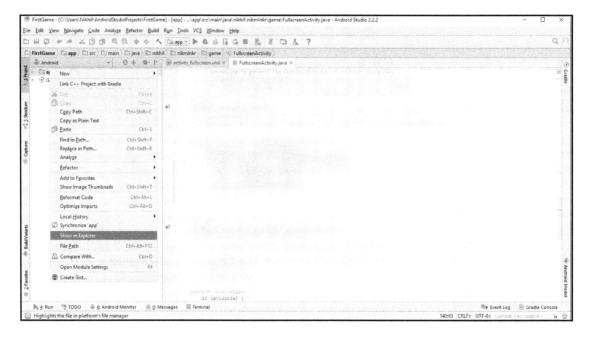

Show in Explorer will open the window in which your project folder is located:

In your project folder, navigate to the `app/build/outputs/apk/` folder, and there you will see your `app-debug.apk` file. Transfer this onto your mobile phone and install the apk and run it:

Click on **Install**:

Click on **Open**:

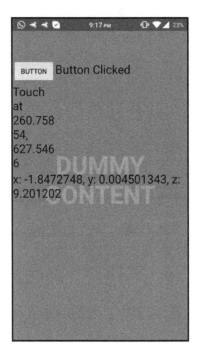

Accelerometer values will change as you move your phone around.

Congratulations! You also learned how to create an apk and run it on your actual device.

Summary

In this chapter, you have learned how to take different types of user inputs and also how to generate an apk and run it on your actual device. You also learned the various types of XML files.

In the next chapter, we will learn how to create sprites--no, not the cold drink! We will learn how to create images as well as play with colors. The next chapter will be our entry into creating graphics on screen and will be a big step for you.

4
Creating Sprites and Interactive Objects

We have learned almost everything about the basics that we need to create various components in Android, so we can now move on to do some more exciting stuff. Now, at this point, we will start working on a proper 2D game. It will be a small 2D side scroller game like Mario. However, before we do that, let's first talk about games as a development concept. In order to understand more about games, you will need to understand a bit of **Game Theory**. So, before we proceed with creating images and backgrounds on screen, let's dive into some game theory. Here are a list of topics we will be covering in this chapter:

- Game Theory
- Working with colors
- Creating images on screen
- Making a continuous scrolling background

Let's start with the first one.

Game Theory

If you observe a game carefully in its source code level, you will observe that a game is just a set of illusions to create certain effects and display them on screen. Perhaps, the best example of this can be the game that we are about to develop. In order to make your character move ahead, you can do either of two things:

- Make the character move ahead
- Make the background move behind

Let's take a look at this in a bit more detail. The preceding two points can be achieved with the help of some illusions; let's understand how.

Illusions

Either of the two things mentioned in the preceding section will give you an illusion that the character is moving in a certain direction. Also, if you remember Mario properly, then you will notice that the clouds and grasses are one and the same, only their colors were changed. This was because of the memory limitations of the console platform at the time:

Game developers use many such *tricks* in order to get their game running. Of course, in today's times, we don't have to worry much about memory limitations because our mobile device has the capability of the Apollo 11 rocket, which landed on the moon. Now, keeping in mind the mentioned two scenarios; we will use one of them in our game to make our character move.

We also have to understand that every game is a loop of activities. Unlike an app, you will need to draw your game resources at every single frame. The illusion of moving or any other effect will be stronger the more frames per second the mobile device can draw them. This concept is called as **Frames Per Second** (**FPS**). It's almost similar to that of the concept of old films where a huge film used to be projected on the screen by rolling per frame. Take a look at the following screenshot to understand this concept better:

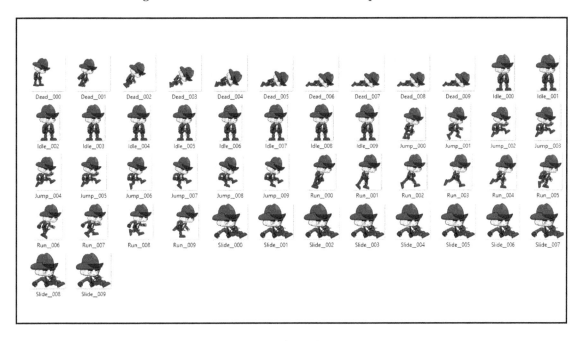

Sprite sheet of a game character

You must have been wondering since the last chapter what a sprite means, if not the popular cold drink. As you can see in the preceding screenshot, a sprite sheet is simply an image consisting of multiple images within themselves in order to create an animation, and thereby a sprite is simply an image. If we want to make our character run, we will simply read the file Run_000 and play it all the way sequentially through to Run_009, which will make it appear as though the character is running. We will take a look at this in *Chapter 5, Adding Animation to Your Game*, which we will follow through.

A majority of things that you will be working with when making a game would be based on manipulating your movements. So, you will need to be clear about your coordinates system because it will come in handy—be it for firing a bullet out of a gun, character movement, or simply turning around to look here and there--all of it is based on the simple component of movement.

Game loop

In its core, every game is basically just a loop of events. It is a set up to give calls to various functions and code blocks to execute in order to have draw calls on your screen, and thereby making the game playable. Mostly, your game comprises three parts:

- Initialize
- Update
- Draw

Initializing the game means to set an entry point to your game through which the other two parts can be called. Your game begins here and is called just once.

Once your game is initialized, you need to start giving calls to your events that can be managed through your `update` function.

The `draw` function is responsible for drawing all your image data on the screen. Everything you see on the screen including your backgrounds, images, or even your GUI is the responsibility of the `draw` method.

To say the least, your game loop is the heart of your game. This is just a basic overview of the game loop, and there is much more complexity you can add to it. However, for now, this much information is sufficient for you to get started.

The following image perfectly illustrates what a game loop is:

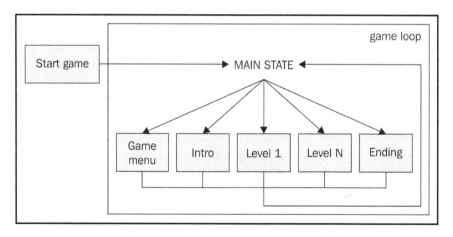

Image Source: `https://gamedevelopment.tutsplus.com/articles/gamedev-glossary-what-is-the-game-loop--gamedev-2469`

Game Design Document

Before starting a game, it is essential to create a **Game Design Document (GDD)**. This document serves as a groundwork for the game you will be making. In all, 99% of the time when we start making a game, we lose track of the features planned for it and deviate from the core game experience. So, it is always recommended to have a GDD in place in order to keep focus. A GDD consists of the following things:

- Gameplay mechanics
- Story (if any)
- Level design
- Sound and music
- UI planning and game controls

You can read more about the Game Design Document by navigating to the following link:

```
https://en.wikipedia.org/wiki/Game_design_document
```

Prototyping

When making a game, we need to test it simultaneously. A game is one of the most complex pieces of software, and if we mess up on one part there is a chance that it might break the entire game as a whole. This process can be called as **Prototyping**. Making a prototype of your game is one of the most important aspects of a game because this is where you test out the basic mechanics of your game. A prototype should be a simple working model of your game with basic functionality. It can also be termed as a **stripped down version** of your game.

Surface and Canvas

We saw how to create images and buttons using components from our palette in an Android app. However, this concept is a bit different in games. In games, we work with something known as a **Canvas**, which is used to draw images onto our **Surface**. To give you a basic understanding, a Surface is anything that holds pixels onto it. Basically, a Surface holds your Canvas, which then maps it onto your views. All of your image manipulations are is based on this. So, for the purpose of drawing anything in our game, we will use a `SurfaceView` component.

Working with colors and images

Now that we have learned these concepts as well as having an understanding of working on Android Studio, we can proceed to start working on making our game from scratch. Let's clear all the buttons and text views in our `activity_fullscreen.xml` file and remove all the references in our Java code so that it looks somewhat like this:

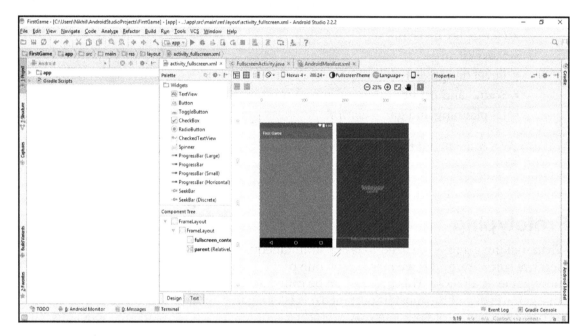

Take a look at the **Component Tree** window below your **Palette** for a proper reference:

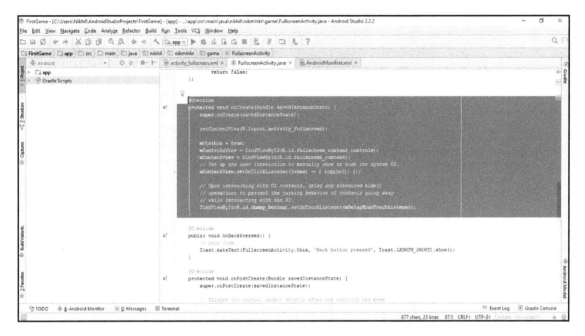

Note here that at this point we are back to square one with our app.

When you are done with this, just run and test your app once to check whether there are any errors; if not, then proceed further. We will now take a look at how to create basic colors using hex color codes and then proceed to create images for our background as well as other components.

Creating colors

This is fairly simple, and we have seen this in our previous chapter as well. Simply go to your `colors.xml` file located in the `app/res/values/` folder. Here, you can observe multiple hex color code values. A hex color code represents a six-digit alphanumeric value, which is responsible for giving a color. The value starts from #**000000** (black) to #**ffffff** (white).

In our `colors.xml`, if you observe there are already predefined values as seen in the following screenshot:

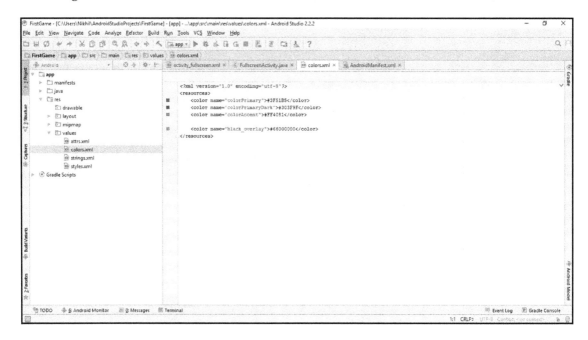

Hex color code values

In this file, you can tweak in these values or simply add your own values as well. Let's say if you want to add the color red, then you simply add the following line into this code to get the red value:

```
<color name="red">#ff0000</color>
```

If you observe the syntax closely, you will observe that this hex code is in the format of `RRGGBB`, which means that the first two alphanumeric digits constitute the composition of the red color, the second two digits correspond to the green color, and the last two digits correspond to that of the blue color. Also, with the name field, you can use this color in any component as we have done in our background. Go ahead, try and tweak some values in the `colorPrimary`, and see what changes happen.

 Also, note here that in hex color code, we can only use numbers from 0-9 and characters from a-f. So, if you use something such as #99z9pt, then this won't give out any color or number output. Try this out so that you get a clearer understanding of this color code concept.

That is all about colors in Android Studio. You can use them in your backgrounds, surfaces, and so on using your understanding of XML files as we have seen in previous chapters, or you also set them programmatically from your Java code. Let's now move on to the most interesting part, working with images.

Creating images

We've taken a look at the project structure of an Android project. So, by now we are clear about the difference between where to put code files and where to put our resources. Images are resources, and therefore, they can either be put in the res folder of our Android Studio Project or AssetManager, but more specifically, we can't just put them in the res folder.

There is, by convention, a specific folder exclusively for our image resources, that is, the drawable folder. You can find this folder in app/res/drawable. If you are not able to see it, then create it manually by right-clicking on the res folder and selecting **New** | **Android resource directory,** and in **Resource Type**, select **drawable**. You do not have to do this if the drawable folder already exists in your folder:

Creating the drawable folder

This folder includes all your image resources in your project including resources, such as background images, icons, and sprite sheets. For the purpose of this game, we'll put all our image resources into the `drawable` folder. We will now create a background image, and in our next chapter, we will learn how to put another image on top of that. After we are done with this, we will create an `OnClickListener` on our newly created image on top of our background so that it changes the image. So for this, we will need three image resources. Let's call them the following:

- `background_image`
- `image_1`
- `image_2`

For this chapter, we have taken the following stock images. You are free to use any images of your choice as you deem fit.

`background_image`: The following is the background image we will be using for our game:

`image_1`: The following is one of the frames of our player:

`image_2`: The following is another image frame of our character:

So, we will have our `background_image` and `image_1` on top of it. When we click on our `image_1`, it will get converted into `image_2`. Let's get some stock image resources and put them into our `res/drawable` folder. To do this, simply right-click on any images you want to put in the `drawable` folder and select **Copy**:

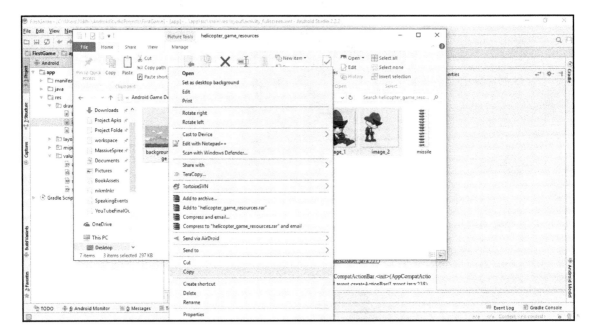

And then, right-click on the **drawable** folder and select **Paste**:

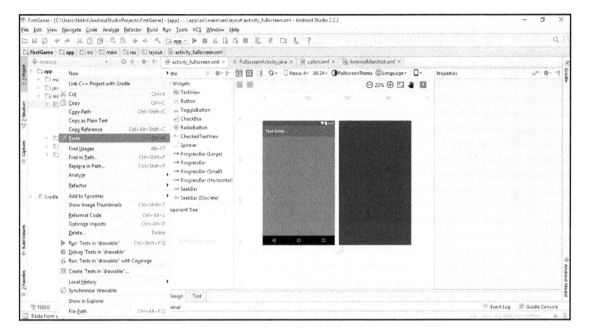

After this, you will be prompted with a dialog box. Press **Ok** to successfully import your image resources in your `project` folder.

Now that we have our image resources in place, it's time to get them on our screen. You can do this right away by taking an **ImageView** component and setting its property to your desired image, but since we are making a game, we will go the *Canvas Way*. To do that, first we need to replace our entire Java code, in our `FullscreenActivity.java` and make it look like this:

```java
package nikhil.nikmlnkr.game;

import android.os.Bundle;
import android.view.Window;
import android.view.WindowManager;
import android.app.Activity;

public class FullscreenActivity extends Activity{
    @Override
    protected void onCreate(Bundle savedInstanceState) {
        super.onCreate(savedInstanceState);
        //Set our game to full screen mode
        getWindow().setFlags
```

```
(WindowManager.LayoutParams.FLAG_FULLSCREEN,
WindowManager.LayoutParams.FLAG_FULLSCREEN);

//Set no title on screen
requestWindowFeature(Window.FEATURE_NO_TITLE);

setContentView(new GameView(this));
    }
}
```

 All the changes you need to do in your existing code have been marked in **bold**

Note here that we have eliminated all the toggle status bar functions and have kept only our `onCreate` method. The other interesting thing we did with the `getWindow().setFlags(WindowManager.LayoutParams.FLAG_FULLSCREEN, WindowManager.LayoutParams.FLAG_FULLSCREEN);` is that we have manually set our game to fullscreen mode. We also eliminated the Title screen window with `requestWindowFeature(Window.FEATURE_NO_TITLE);` code. Now, note here you will be getting an error on your `GameView(this)` code. This is because we still haven't created our `GameView` class. So, let's go ahead and make that but before that let's make one minor change in our manifest file.

Open up your `AndroidManifest.xml` file located in the `app/manifests/` folder. In your `<activity>` code, add the orientation as landscape as follows:

```xml
<?xml version="1.0" encoding="utf-8"?>
<manifest xmlns:android="http://schemas.android.com/apk/res/android"
    package="nikhil.nikmlnkr.game">

    <application
        android:allowBackup="true"
        android:icon="@mipmap/ic_launcher"
        android:label="@string/app_name"
        android:supportsRtl="true"
        android:theme="@style/AppTheme">
        <activity
            android:screenOrientation="landscape"
            android:name=".FullscreenActivity"
            android:configChanges="orientation
            |keyboardHidden|screenSize"
            android:label="@string/app_name"
            android:theme="@style/FullscreenTheme">
```

```
        <intent-filter>
            <action android:name="android.intent.action.MAIN" />

            <category android:name=
            "android.intent.category.LAUNCHER" />
        </intent-filter>
    </activity>
</application>

</manifest>
```

This will now explicitly tell the application that our game is in landscape mode.

Now, let's move ahead and make our `GameView` class. To do so, simply right-click on your `app/java/packagename` folder and select **New** | **Java Class**, like this:

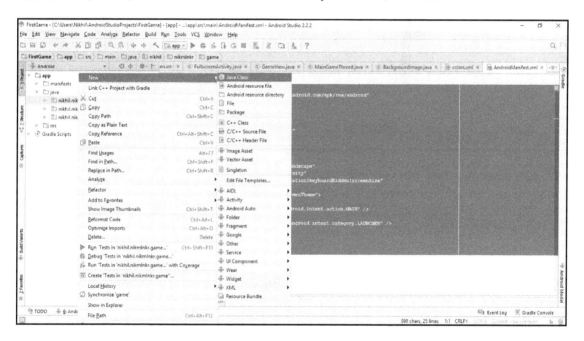

After you do so, a new window will open up asking you the details of the new class you want to make. Simply type in GameView inside of the **Name** text box and press **OK** to proceed, as shown in the following screenshot:

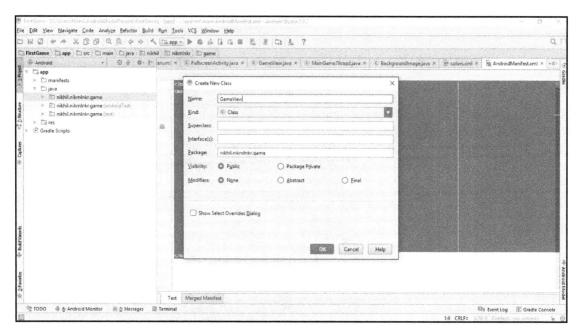

We will be needing two more classes to set our side-scrolling background image, so repeat the same process and create the following two classes:

- BackgroundImage
- MainGameThread

So, in total, you now have four classes in your project:

- BackgroundImage
- FullscreenActivity
- GameView
- MainGameThread

Our aim is to have a background image on the screen, which would continuously scroll through our view. Let's first open up our BackgroundImage.java file and write the following code in it:

```java
package nikhil.nikmlnkr.game;

import android.graphics.Bitmap;
import android.graphics.Canvas;

public class BackgroundImage {

    private int xc, yc, dxc;
    private Bitmap backgroundImage;

    public BackgroundImage(Bitmap res)
    {
        backgroundImage = res;
    }

    public void setVector(int dxc)
    {
        this.dxc = dxc;
    }

    public void update()
    {
        xc += dxc;
        if(xc < -GameView.WIDTH){
            xc=0;
        }
    }

    public void draw(Canvas canvas)
    {
        canvas.drawBitmap(backgroundImage, xc, yc,null);
        if(xc < 0)
        {
            canvas.drawBitmap
            (backgroundImage, xc + GameView.WIDTH, yc, null);
        }
    }
}
```

Let's try to understand this code now. We have simply created a class here and following is a step-by-step breakdown of each code block inside our class:

1. We imported Bitmap and Canvas, which is needed for our image and Canvas operations.
2. We declared private variables xc, yc, and dxc, which are simply the *x, y* coordinates and displacement in *x*. The default value of these variables will be zero since we haven't initialized them.
3. We then declared a Bitmap private variable, which will be holding our actual image file to be displayed on the screen.
4. Next, we created a constructor for our class, so we can pass an image into it using the res reference inside it and within our constructor, we equated this reference with our background image variable declared in step 3.
5. We then created a vector displacement method in order to add a unit vector to our image's position value in order to make it move.
6. Then, we used the update method that will be called every time and put all our displacement and reset logic in it. If the image goes out of our screen, then we reset the position to 0 in order to give a continuous movement effect.
7. Now, if you remember, we discussed at the beginning that in order to draw anything on the screen, you need a Canvas. So, using the draw method, we took a Canvas as a reference variable and included our draw logic in it. Observe here that we are drawing our background image twice. This is because if our image is scrolling continuously, then there will be a void in between and it'll appear black, so we use the same image and draw it twice on screen a bit further from our main image in order to give a continuous loop effect. In order to understand this as a live example, you can try removing either one of the `canvas.drawBitmap()` code to see for yourself.

That's it for the `BackgroundImage.java` file. By this time, you must be getting a few errors on your `GameView.WIDTH` code. Don't worry about that. We will come to it later. Before addressing that error, we must first set up our game thread since all our update functions are going to be called from our thread. Our objectives for our `MainGameThread.java` file are the following:

- Give a continuous call to the update function
- Base our performance on frames per second

So, open up your MainGameThread.java file and type in the following:

```java
package nikhil.nikmlnkr.game;
import android.graphics.Canvas;
import android.view.SurfaceHolder;

public class MainGameThread extends Thread
{
    private int framesPerSecond = 30;
    private double averageFPS;
    private SurfaceHolder surfaceHolder;
    private GameView gameView;
    private boolean running;
    public static Canvas canvas;

    public MainGameThread(SurfaceHolder surfaceHolder, GameView
    gameView){
        super();
        this.surfaceHolder = surfaceHolder;
        this.gameView = gameView;
    }

    public void setRunning(boolean b){
        running=b;
    }

    @Override
    public void run(){
        long startTime;
        long timeMillis;
        long waitTime;
        long totalTime = 0;
        int frameCount =0;
        long targetTime = 1000/framesPerSecond;

        while(running) {
            startTime = System.nanoTime();
            canvas = null;

            //try locking the canvas for pixel editing
            try {
                canvas = this.surfaceHolder.lockCanvas();
                synchronized (surfaceHolder) {
                    this.gameView.update();
                    this.gameView.draw(canvas);
                }
            } catch (Exception e) {
            }
```

```
finally{
    if(canvas!=null)
    {
        try {
            surfaceHolder.unlockCanvasAndPost(canvas);
        }
        catch(Exception e){e.printStackTrace();}
    }
}
timeMillis = (System.nanoTime() - startTime) / 1000000;
waitTime = targetTime-timeMillis;

try{
    this.sleep(waitTime);
}catch(Exception e){}

totalTime += System.nanoTime()-startTime;
frameCount++;
if(frameCount == framesPerSecond){
    averageFPS = 1000/((totalTime/frameCount)/1000000);
    frameCount = 0;
    totalTime = 0;
    System.out.println(averageFPS);
}
        }
    }
}
```

Here's what we have done in our `MainGameThread.java` file:

1. We created all the variables needed to run our thread.
2. We created the constructor for our `MainGameThread` file by taking a `SurfaceHolder` and `GameView` references and set their references to `this` file.
3. We created a method to keep a track of our running thread with a return value of Boolean.
4. We override the default `run` method of a thread to make it do the stuff we need specifically:
    1. We use our surfaceholder variable for manipulating our image pixel data.
    2. We calculate our frames per second.
    3. We take an average of frames per second to calculate and display it in the console view.

By doing this, we are now ready with our thread class and finally only left with our GameView class. In our GameView, we are actually going to put all the image data we built on screen and display it. So, let's open up the GameView.java file and start typing the following:

```
package nikhil.nikmlnkr.game;

import android.content.Context;
import android.graphics.BitmapFactory;
import android.graphics.Canvas;
import android.view.MotionEvent;
import android.view.SurfaceHolder;
import android.view.SurfaceView;


public class GameView extends SurfaceView implements SurfaceHolder.Callback
{
    public static final int WIDTH = 1920;
    public static final int HEIGHT = 1080;
    private MainGameThread mainThread;
    private BackgroundImage bgImg;

    public GameView(Context context){
        super(context);


        //set callback to the surfaceholder to track events
        getHolder().addCallback(this);

        mainThread = new MainGameThread(getHolder(), this);

        //make gamePanel focusable so it can handle events
        setFocusable(true);
    }

    @Override
    public void surfaceChanged(SurfaceHolder holder, int format,
    int width, int height){}

    @Override
    public void surfaceDestroyed(SurfaceHolder holder){
        boolean retry = true;
        while(retry){
            try{
                mainThread.setRunning(false);
                mainThread.join();
            }catch(InterruptedException e){e.printStackTrace();}
```

```
                    retry = false;
            }

    }

    @Override
    public void surfaceCreated(SurfaceHolder holder){

        bgImg = new BackgroundImage(BitmapFactory.decodeResource
        (getResources(), R.drawable.background_image));
        bgImg.setVector(-5);

        //we can safely start the game loop
        mainThread.setRunning(true);
        mainThread.start();

    }
    @Override
    public boolean onTouchEvent(MotionEvent event){
        return super.onTouchEvent(event);
    }

    public void update(){
        bgImg.update();
    }

    @Override
    public void draw(Canvas canvas){
        final float scaleFactorX = getWidth()/WIDTH;
        final float scaleFactorY = getHeight()/HEIGHT;
        if(canvas != null) {
            final int savedState = canvas.save();
            canvas.scale(scaleFactorX, scaleFactorY);
            bgImg.draw(canvas);
            canvas.restoreToCount(savedState);
        }
    }
}
```

This code is pretty easy to understand. Our GameView is the place where we create our Surface and draw everything onto it. So, we have extended the SurfaceView and implemented our callback, SurfaceHolder. This gives us access to some pre-written methods, which we will be overriding. Before you understand these methods, let's try to understand the logic behind this code. It can be divided into the following:

1. We create the default constructor for our GameView class, which then gives a call to start our MainGameThread.
2. Using predefined methods, we override them and create our Canvas on top of our surface.
3. We call the update function in our BackgroundImage class.
4. We set scaling of our image to match our phone's resolution dynamically.

Now that we know the logic, we can now read through the methods and understand them each one by one. We have the following methods in our code:

- surfaceChanged: We have created an empty method with parameters. If our surface changes, then this method is called
- surfaceDestroyed: If the surface is destroyed, this method is called
- surfaceCreated: After our surface is created, we can start the game loop; this is where we initialize our background image and set its resource, as you can see in the code, bgImg = newBackgroundImage(BitmapFactory.decodeResource(getResources(), R.drawable.background_image));
- onTouchEvent: This method is called whenever there is a touch on the screen
- update: This method is the update method, and in here, we are giving a call to the update method of BackgroundImage
- draw: This method gives a call to draw our image onto the screen, and we make some calculations to scale our image and set it properly onto our phone as per its resolution

After you are done with this file, don't forget to check your `FullscreenActivity.java` and make sure that its code looks like this:

```java
package nikhil.nikmlnkr.game;

import android.os.Bundle;
import android.view.View;
import android.view.Window;
import android.view.WindowManager;
import android.app.Activity;

public class FullscreenActivity extends Activity{

    @Override
    protected void onCreate(Bundle savedInstanceState) {
        super.onCreate(savedInstanceState);
        //Set our game to full screen mode
        getWindow().setFlags
        (WindowManager.LayoutParams.FLAG_FULLSCREEN,
        WindowManager.LayoutParams.FLAG_FULLSCREEN);

        //Set no title on screen
        requestWindowFeature(Window.FEATURE_NO_TITLE);

        setContentView(new GameView(this));
    }
}
```

 It is now recommended that you build an apk for this game and test it on your device since if you try running this on your emulator it will run extremely slow.

Build it and execute it on your device/emulator, and you will see an output like this in which your background image is continuously scrolling:

That's it! Your side-scrolling background is up and running.

Summary

Congratulations! You have successfully learned how to create images and work with colors in Android Studio. You have also implemented a side-scrolling background, which will act as a foundation for the further part of this game.

In the next chapter, we will create our player character and implement a click listener on our image objects as well as learn how to animate objects with the concept of sprite sheets we have just learned.

5
Adding Animation to Your Game

We have learned how to create sprites on screen as well as made our background image scroll continuously. Sprites are basically just images, which are used for game assets. Now, it is time to add a bit more spice to it and make it interesting. In this chapter, we will make our player character appear on screen on top of our background and add an animation that will make the player run. In this chapter, you will learn the following:

- Creating sprite animations based on sprite sheets
- Running basic animations
- Creating an abstract class to serve as a foundation for our future game objects

By the end of this chapter, you will have a running character on screen. So let's begin!

Adding animations to make your game more awesome

To start this chapter, we will need a collection of images, which we call sprites. We saw in the previous chapter how we had used `image_1` and `image_2` as examples, but we will expand on them and make this more interesting. We will try and make a proper run cycle of our player character in a sprite sheet. To understand this in a better way, let's take a look at the following image:

Our run cycle animation sprites

In the preceding image, you can observe we have a set of images that are prefixed with the text Run followed by a set of numbers that represent frames. So, basically, we will loop from frame `Run_000` to `Run_009`, which will give us a continuous running effect. However, for the sake of simplicity, we will be merging these frames into one single image and read pixel data from it. Also, to keep it simple, we will only make the run cycle animation to three frames. If you want, you can add more frames for better animation quality. So, we will be using the following sprite sheet:

This is what an actual sprite sheet looks like; we will name this sprite sheet as **player_run.png**

We will be dealing with pixels to run our animation, so it is extremely important for us to know the dimensions of this image. In simple terms, we need to know the width and height of this image in pixels. The dimensions of one frame of our image are 200 x 82 pixels, where 200 represents the width and 82 represents the height of our image. In our sprite sheet, we have three such sprites. Thus, we simply multiply our width by the number of images in our sprite sheet, which is three in this case. So, the overall width of our image is 600 pixels. As you can see, we are using an image that has a landscape orientation, which means that our image's width is more than the height, so our height dimension remains the same. So, basically to run our animation, we only need to scan through our frames horizontally. The final resolution of our sprite sheet is 200 x 82.

 The resolution of your image might differ from the ones in this book, so make sure that you calculate the numbers properly before proceeding with writing them in the code.

So, let's dive into the art of animating our player character; however, before we do that, we also need to create our player character on screen. Here's our task list for now:

1. Create a player on screen.
2. Make Run animation play.

Once we are done with these two objectives, we will take touch input in order to make our player jump.

Making our player character

We saw in the previous chapter how to create our background image. In theory, it is almost similar to the previous chapter to create our player character, but since we are going to deal with animations in the further part of this chapter, we will need to modify our code a bit. Let's get started with that first.

Now, as a programmer, you have to keep in mind that there are many ways to do one particular thing; therefore, just as a demonstration to get a clearer understanding, we will modify our background image code a little. Open up your BackgroundImage.java file and remove the setVector method. Now, you will see that there will be an error in our GameView.java file since our setVector method does not exist; let's fix that. Make a static final variable, which will be accessible from any class. We will then set it as a speed variable. Then, in our BackgroundImage.java file's constructor, we will set the displacement variable as the speed for this. Here's how we will modify our code blocks. Code changes are marked in bold.

The following is the code for BackgroundImage.java:

```
package nikhil.nikmlnkr.game;

import android.graphics.Bitmap;
import android.graphics.Canvas;

public class BackgroundImage {

    private int xc, yc, dxc;
    private Bitmap backgroundImage;

    public BackgroundImage(Bitmap res){
        backgroundImage = res;
        dxc = GameView.MOVINGSPEED;
    }


    public void update(){
        xc += dxc;
        if(xc < -GameView.WIDTH){
            xc=0;
        }
    }

    public void draw(Canvas canvas){
        canvas.drawBitmap(backgroundImage, xc, yc,null);
        if(xc < 0){
            canvas.drawBitmap(backgroundImage, xc + GameView.WIDTH, yc,
```

```
null);
        }
    }
}
```

Next, let's take a look at the code for GameView.java:

```java
package nikhil.nikmlnkr.game;

import android.content.Context;
import android.graphics.BitmapFactory;
import android.graphics.Canvas;
import android.view.MotionEvent;
import android.view.SurfaceHolder;
import android.view.SurfaceView;


public class GameView extends SurfaceView implements SurfaceHolder.Callback
{
    public static final int WIDTH = 1920;
    public static final int HEIGHT = 1080;
    public static final int MOVINGSPEED = -5;
    private MainGameThread mainThread;
    private BackgroundImage bgImg;

    public GameView(Context context){
    super(context);


        //set callback to the surfaceholder to track events
        getHolder().addCallback(this);

        mainThread = new MainGameThread(getHolder(), this);

        //make gamePanel focusable so it can handle events
        setFocusable(true);
    }

    @Override
    public void surfaceChanged(SurfaceHolder holder, int format,
    int width, int height){}

    @Override
    public void surfaceDestroyed(SurfaceHolder holder){
        boolean retry = true;
        while(retry){
            try{
                mainThread.setRunning(false);
```

```
                mainThread.join();
            }catch(InterruptedException e){e.printStackTrace();
            }
            retry = false;
        }
    }

    @Override
    public void surfaceCreated(SurfaceHolder holder){

        bgImg = new BackgroundImage(BitmapFactory.decodeResource
        (getResources(), R.drawable.background_image));

        //we can safely start the game loop
        mainThread.setRunning(true);
        mainThread.start();
    }

    @Override
    public boolean onTouchEvent(MotionEvent event){
        return super.onTouchEvent(event);
    }

    public void update(){
        bgImg.update();
    }

    @Override
    public void draw(Canvas canvas){
        final float scaleFactorX = getWidth()/WIDTH;
        final float scaleFactorY = getHeight()/HEIGHT;
        if(canvas != null) {
            final int savedState = canvas.save();
            canvas.scale(scaleFactorX, scaleFactorY);
            bgImg.draw(canvas);
            canvas.restoreToCount(savedState);
        }
    }
}
```

Now, we haven't done anything really here. We have just modified a logic with different variables. In a similar fashion, you can also apply your own logic to come up with different ways to do one particular thing. Let's now move on to actually make our player character appear on the screen.

Try tweaking the code with your own logic and keeping the output same as a challenge to test your skills.

We will now create an abstract class for our future game objects, such as our player character, missiles, and everything else. We are doing this because there are some sets of data, which are required in almost every object we will be creating in future for our game. So, for the purpose of reusability and persistence, we will create this class. We will call it GameObj.java. Go ahead and create your new class with the steps that we learned from our previous chapters and write the following code in it:

```java
package nikhil.nikmlnkr.game;

import android.graphics.Rect;

/**
 * Created by Nikhil on 13-01-2017.
 */

public abstract class GameObj {
    protected int xc, yc, dxc, dyc;
    //Our x and y coordinates along with their displacement variables
    protected int width, height;
    //width and height of our objects

    public int getXC() {
        return xc;
    }

    public int getYC() {
        return yc;
    }

    public void setXC(int xc) {
        this.xc = xc;
    }

    public void setYC(int yc) {
        this.yc = yc;
    }

    public int getWidth() {
        return width;
    }

    public int getHeight() {
```

```
        return height;
    }

    public Rect getRectangle() {
        return new Rect(xc, yc, xc + width, yc + height);
    }
}
```

So, here we have our abstract GameObj.java class file, which has the variables and get set methods for them. Observe here that we have also used a getRectangle() method. This method will be used in the next chapter when we will be working with collisions. In theory, to detect collisions on any object, we need to get its rectangular bounds. Anyway, moving ahead, let's now create our Player class with the help of this new abstract class we made.

Create a new class and name it PlayerCharacter.java, and write the following code in it:

```
package nikhil.nikmlnkr.game;

import android.graphics.Bitmap;
import android.graphics.Canvas;

/**
 * Created by Nikhil on 13-01-2017.
 */

public class PlayerCharacter extends GameObj{
private Bitmap spriteSheet;
private int score;
private double dya;
private boolean up, playing;
private AnimationClass ac = new AnimationClass();
private long startTime;

    public PlayerCharacter(Bitmap res, int w, int h, int noOfFrames) {
        xc = 100;
        yc = GameView.HEIGHT/2;
        dyc = 0;
        score = 0;
        height = h;
        width = w;

        Bitmap[] img = new Bitmap[noOfFrames];
        spriteSheet = res;

        for(int i=0; i < img.length;i++){
                img[i] = Bitmap.createBitmap(spriteSheet, i*width, 0,
```

```
width, height);
        }
        ac.setFrames(img);
        ac.setDelay(10);
        startTime = System.nanoTime();
    }

    public void setUp(boolean b){
        up = b;
    }

    public void update() {
        long elapsed = (System.nanoTime()-startTime)/1000000;
        if(elapsed > 100) {
            score++;
            startTime = System.nanoTime();
        }
        ac.update();
    }

    public void draw(Canvas canvas) {
        canvas.drawBitmap(ac.getImage(), xc, yc, null);
    }

    public int getScore() {
        return score;
    }

    public boolean getPlaying(){
        return playing;
    }

    public void setPlaying(boolean b) {
        playing = b;
    }

    public void resetDYA() {
        dya = 0;
    }

    public void resetScore () {
        score = 0;
    }
}
```

Let's now understand what we have done in our `PlayerCharacter` code in further detail:

- We created our class and extended it to our abstract class, `GameObj`, to get access to all the default variables needed for our `PlayerCharacter`
- We created the default constructor for our `PlayerCharacter` in which we pass the required data to draw `PlayerCharacter` on screen just like the image component is passed in the `res` variable along with the width and height of our image with the number of frames we need in our animation
- In our constructor, we created a `for` loop, which will run through our sprite sheet and give us the animation effect followed by a 10-millisecond delay to play our animations
- We then created our `setUp()` function, which will take care of the jumping functionality of our player
- In our `update()` function, we created a loop of events that assign a `score` to the player as well as keep the player between upper and lower bounds
- After this, we created the `draw()` method, which gets the animation from our `ac` variable and drew it onto our canvas
- This is then followed by simple get and set methods for rest of the variables

Our player character is ready; however, we still have to write our `AnimationClass`. As you can clearly observe, you must be getting an error on the `AnimationClass` line. So, let's go ahead and create our `AnimationClass.java` file and write the following code in it:

```
package nikhil.nikmlnkr.game;

import android.graphics.Bitmap;

/**
 * Created by Nikhil on 13-01-2017.
 */

public class AnimationClass {
    private Bitmap[] frames;
    private int currentFrame;
    private long startTime, delay;
    private boolean playedOnce;

    public void setFrames(Bitmap[] frames){
        this.frames = frames;
        currentFrame = 0;
        startTime = System.nanoTime();
    }
```

```
public void setDelay(long d){
    delay = d;
}

public void setFrame(int i) {
    currentFrame = i;
}

public void update() {
    long elapsed = (System.nanoTime()-startTime)/1000000;

    if(elapsed > delay) {
        currentFrame++;
        startTime = System.nanoTime();
    }

    if(currentFrame == frames.length) {
        currentFrame = 0;
        playedOnce = true;
    }
}

public Bitmap getImage(){
    return frames[currentFrame];
}

public int getFrame(){
    return currentFrame;
}

public boolean playedOnce() {
    return playedOnce;
}
}
```

This AnimationClass.java file is important to run our character's animations. Our AnimationClass has three main functions, namely setFrames(), update(), and getImage(). Let's take a look at what we have done in this class:

- We created variables that are needed to access our frames
- We created variables that will control the start time and delay between animation frames
- We created a Boolean, which will trigger the animation
- Next, we created a setFrames() function, to set frames and set the current frame to 0 at start

- After that, a `setDelay()` function, was created in order to tell the animation to run at short intervals
- We then created a `setFrame()` function to set the current frame in the running animation
- In our `update()` function, we started calculating the delay in terms of milliseconds using some simple math calculations, and upon certain intervals, switched the frame of the image
- We added a condition that if our last frame has elapsed then we will reset our current frame to 0 in order to give a continuous looping effect
- In our `getImage()` function, we simply return the value of the image that is currently being drawn on the screen
- The `getFrame()` function returns us the integer value of the current frame that is being displayed
- With the last function, `playedOnce()`, as a placeholder for our further gameplay

setFrames() and setFrame() are two different functions.
setFrames() is used to initialize our frames from 0, whereas
setFrame() is used to set frames individually while running.

Now, before moving on to the next part, ensure that you have your image in your drawable folder that you want to use for our player character. Once you have it, it's now time to move on to the next step and instantiate our player on screen. The procedure for this is the same as we did with our background image. Modify your `GameView.java` file code to look like this; changes from the preceding code have been highlighted in bold:

```java
package nikhil.nikmlnkr.game;

import android.content.Context;
import android.graphics.BitmapFactory;
import android.graphics.Canvas;
import android.view.MotionEvent;
import android.view.SurfaceHolder;
import android.view.SurfaceView;


public class GameView extends SurfaceView implements SurfaceHolder.Callback
{
    public static final int WIDTH = 1920;
    public static final int HEIGHT = 1080;
    public static final int MOVINGSPEED = -5;
    private MainGameThread mainThread;
    private BackgroundImage bgImg;
```

```
private PlayerCharacter playerCharacter;

public GameView(Context context) {
    super(context);


    //set callback to the surfaceholder to track events
    getHolder().addCallback(this);

    mainThread = new MainGameThread(getHolder(), this);

    //make gamePanel focusable so it can handle events
    setFocusable(true);
}

@Override
public void surfaceChanged
(SurfaceHolder holder, int format, int width, int height){}

@Override
public void surfaceDestroyed(SurfaceHolder holder){
boolean retry = true;
while(retry)
    {
        try{mainThread.setRunning(false);
            mainThread.join();

    }catch(InterruptedException e){e.printStackTrace();}
        retry = false;
    }


}

@Override
public void surfaceCreated(SurfaceHolder holder){

bgImg = new
BackgroundImage(BitmapFactory.decodeResource
    (getResources(),
    R.drawable.background_image));
    Drawable d = getResources().getDrawable
    (R.drawable.player_run);
    int w = d.getIntrinsicWidth();
    int h = d.getIntrinsicHeight();
    playerCharacter = new PlayerCharacter
    (BitmapFactory.decodeResource
    (getResources(),R.drawable.player_run),w/3,h,3);
    //we can safely start the game loop
```

```
        mainThread.setRunning(true);
        mainThread.start();

    }
    @Override
    public boolean onTouchEvent(MotionEvent event){
        return super.onTouchEvent(event);
    }

    public void update(){
        bgImg.update();
        playerCharacter.update();
    }
    @Override
    public void draw(Canvas canvas){
        final float scaleFactorX = getWidth()/WIDTH;
        final float scaleFactorY = getHeight()/HEIGHT;
        if(canvas!=null) {
            final int savedState = canvas.save();
            canvas.scale(scaleFactorX, scaleFactorY);
            bgImg.draw(canvas);
            playerCharacter.draw(canvas);
            canvas.restoreToCount(savedState);
        }
    }

}
```

Let's do a breakdown of what we did here:

1. We created a reference variable for our `PlayerCharacter` using the variable `playerCharacter`.

2. Then, in our `surfaceCreated()` method, we gave the `playerCharacter` class constructor all the values that it requires. We passed the `player_run` image into this code for it to fetch our player's sprite sheet. In the parameters for the constructor of `PlayerCharacter(BitmapFactory.decodeResource(getResources(),R.d rawable.player_run),w/3,h,3);`, the 'w' corresponds to the width of the image and 'h' corresponds to the height of the image. The parameter 3 here will depend on the number of frames of your sprite. If you have six frames in your sprite, then the parameters would be `(w/6,h,6)`.

3. After that, in our `update()` method, which is after `bgImg.update()`, we called the `playerCharacter.update()` method that gives a call to the update function of our `PlayerCharacter` and thereby plays the animation from the `AnimationClass.java` file.

4. Lastly, we then drew our player character onto our canvas using the `playerCharacter.draw(canvas);` code.

After you are done with these steps, build and run your game on an emulator or your Android phone device; you will get the following output:

Hurray! Our player character is running on the screen (literally)

If you have done everything correctly, then you will have your player character running on the screen on the spot while your background is scrolling continuously; almost half of our game is done here.

Since we have created our `GameObj` class, we will be relying on it heavily for our further game objects, such as missiles, particle effects, and everything else.

Congratulations! You just created your first animation!

Summary

In this chapter, you learned how to effectively create animations from your sprite sheet as well as create an abstract class to create a foundation for your future game objects. In the next chapter, we will learn how to create our game even more interesting by controlling our animation based on touch inputs. We will learn about collision detection along with creating a missile AI that will spawn from the extreme right of the screen, and our objective will be to dodge the missile.

We will learn about different collision techniques as well as create interactive objects, where we will also add a score on collecting them instead of just incrementing the score with time.

6
Collision Detection and Basic Artificial Intelligence

You learned an overview on how to play animations in our game, and now we are going to go a step further into this exciting journey of game development by learning one of the most complex yet crucial concepts required to make our game feel realistic. By adding animations, we can make our game look realistic, but it is also important for a game to feel realistic because that is what makes it fun to play. In this chapter, we will be taking a look at the following two concepts in further detail:

- Collision detection
- Artificial intelligence

Thus, this chapter will be divided into two main sections. The studies of collision detection and artificial intelligence are in themselves quite vast topics. For the sake of simplicity to serve our purpose, we will be taking a look at the most basic version of these topics to get an entry level understanding for being able to apply the knowledge of these concepts in our game. So without further ado, let's dive into this complex yet exciting topic of collision detection.

Collision detection

To explain it simply, collision is a short duration interaction between two bodies. There are many different types of collisions such as elastic and non-elastic. The study of the intersection of two or more objects overlapping each other is called as **collision detection**. It is one of the most complex pieces of computational mathematics and is divided into many types, such as:

- **Bounding Box Collision:** This is the simplest form of collision techniques wherein we take two rectangles and check for collision if they overlap each other. For this, we need four coordinates of each rectangle, namely, the x and y position and width and height of both rectangles.
- **Circle Collision:** This is the second simplest type of collision wherein we test for collisions between two circles. Here, the radius of two circles and x and y position of the center of the circles are taken into account to test for overlapping.
- **Separating Axis Theorem:** This type of collision is a bit more complex than the other two mentioned earlier, primarily because this is used to detect collisions between two polygons.

There are, of course, many other types of collisions, but that is a totally huge concept in itself. For the purpose of this book, we will be dealing with the simpler forms of collision, and based on the understanding of these concepts, you can experiment further with more complex collision types.

Now, let's look at the algorithms for the three types of collisions you learned about.

 These algorithms are just pseudo codes. Pseudo means false, which means these codes are not to be executed. They are just mentioned here for the purpose of understanding the different collision techniques.

Let's take a look at each of them individually before resuming work on our game project.

Algorithms for collision detection techniques

It is very essential that we get a foundation for understanding these collision techniques and therefore it is very important that we understand the way these techniques work. Let's take a look at the different algorithms for detecting collisions. We will be using these concepts in our game when we are dealing with collisions further in this chapter. Since we will only be working with the simple types in this book, we will only take a look at the algorithms for Bounding Box and Circle Collision Detection techniques.

Bounding Box Collision

We saw in our preceding explanation points that Bounding Box Collision technique is one of the simplest. That's because we are simply testing for between two rectangles. Consider the following pseudo code for a better understanding:

```
rectangle1 = {x: 5, y: 5, width: 50, height: 50}
rectangle2 = {x: 20, y: 10, width: 10, height: 10}

if(rectangle1.x < rectangle2.x + rectangle2.width && rectangle1.width >
rectangle2.x && rectangle1.y < rectangle2.y + rectangle2.height &&
rectangle1.height + rectangle1.y > rectangle2.y)
{
    //Bounding Box Collision Detected
}

// Taking the values from our variables
if (5 < 30 && 55 > 20 && 5 < 20 && 55 > 10) {
 // Bounding Box Collision Detected!
}
```

As you can see from the preceding code, the mathematical operations required to detect a collision through this technique are quite simplistic. We are only working with basic x, y coordinates and width and height.

Circle Collision

Yet another simple type of collision, this deals with mapping out the distance between the centers of two circles to detect a collision. Its algorithm goes as follows:

```
circle1 = {radius: 20, x: 5, y: 5};
circle2 = {radius: 12, x: 10, y: 5};

dx = circle1.x - circle2.x;
dy = circle1.y - circle2.y;
```

```
distance = Math.sqrt(dx * dx + dy * dy);

if (distance < circle1.radius + circle2.radius) {
 // Circle Collision!
}
```

As you can observe here, we have two circles. We then take their individual distances of x and y coordinates. After that, we take the square root of the sum of their squares. This is just the simple formula to calculate the distance between two circles. Then in our testing condition, we check if the distance is less than the sum of two circles.

Now that we have a basic understanding of the algorithms for collision detection, we are equipped to move ahead on detecting the collision in our game. Let's start with it!

Detecting collisions in our game

Since we are dealing with a game here, we will be dealing with the theory with a practical application. So, let's a take up a task of studying collision techniques with an example of an approaching rock toward the player. Our game is a side scrolling game and therefore there will be multiple obstacles and collectibles heading our way. Through the simple Bounding Box Collision technique, we can detect collisions between our player and other objects and execute corresponding functions.

After we are done creating our incoming rocks, we will also create incoming coins so that the player can score points as well. However, before we proceed with this, we have to come up with a way for our player to avoid these obstacles too. We will be implementing collision detection wherein if the player collides with a rock then he will die. Thus, it is crucial for our player to jump. We will give our player jumping abilities on our touch input. So, let's get our player to jump!

Making our player jump

Since in our further chapters, we will be spawning obstacles and coins randomly on the screen, we will modify our player to navigate up and down on the screen. So, let's say if you touch your screen, the player will go up and if you stop touching the screen, our player will go down. Let us take a look at the changes we need to implement in our code first and then understand them step by step by breaking them down. We will be working on the GameView.java file for this code portion.

Initially, our animation used to start as soon as the game started. But we need a little bit more control over our actions so we used our `getPlaying()` function to achieve this. Here's how we will tackle this problem:

We added a condition in our `update()` method that if our player is playing only then will we update our background image and player character. It simply means that the game won't start without some signal:

```
public void update(){
    if(playerCharacter.getPlaying()) {
        bgImg.update();
        playerCharacter.update();
    }
}
```

We will now use the `onTouchEvent()` to notify that our game has started as well as when to go up and when to go down. In the first condition inside our `onTouchEvent()`, we will check if we have a touch event on the screen. `ACTION_DOWN` means that the screen is touched.

Inside this, if blocked, we will have another if block, which checks if the player is playing or not. If the player is not playing, then we set the `setPlaying()` function to set a `true` value, thereby starting the game loop which then starts the `update()` method because of our condition in step 1. Otherwise, it will simply tell the player that the screen is touched and therefore our `up` Boolean variable is set to `true`, which means the player goes up.

After this, we will write the `return true` statement, which is responsible for notifying about the touch event since our `onTouchEvent()` has a Boolean return type.

Next, we define our condition for the player to go down. In our case, this simply means that the touch input is no longer received, which means our finger was lifted from the screen. `ACTION_UP` defines this event.

If this event takes place, then we set our `up` variable to `false` and therefore our player goes down.

We will again write a `return true` statement for our event. Here's what our code for `onTouchEvent()` will look like:

```
@Override
public boolean onTouchEvent(MotionEvent event)
{
    if(event.getAction() == MotionEvent.ACTION_DOWN) {
        if(!playerCharacter.getPlaying()){
            playerCharacter.setPlaying(true);
```

```
            } else {
                playerCharacter.setUp(true);
            }
            return true;
        }

        if(event.getAction() == MotionEvent.ACTION_UP){
            playerCharacter.setUp(false);
            return true;
        }
        return super.onTouchEvent(event);
    }
}
```

From this edit, we have only defined the calls required for setting the up and down functionality of our player. However, we also need to add acceleration and deceleration to our player in order to actually move up and down. We will achieve this by editing our `PlayerCharacter.java` file. Let's write some jumping code in this file. Open it up and write the following code marked in bold in the `update()` method:

```
public void update() {
    long elapsed = (System.nanoTime()-startTime)/1000000;
    if(elapsed > 100) {
        score++;
        startTime = System.nanoTime();
    }
    ac.update();

    if(up){
        dyc = (int)(dya-=1.1);
    }
    else {
        dyc = (int)(dya+=1.1);
    }

    if(dyc > 10) {
        dyc = 10;
    }
    if(dyc < -10) {
        dyc = -10;
    }

    yc += dyc*2;
    dyc = 0;
}
```

This block of code is extremely easy to understand. Let's break it down to get a proper grasp of the same:

1. We check if our `up` variable is true or not.
2. If it is `true` , then we add a positive acceleration value which *increases by 1.1* in every update. You can tweak these values as per your liking.
3. If it is `false`, then we add a negative acceleration value which *decreases by 1.1* in every update.
4. After this, we don't want the acceleration or deceleration to go beyond a certain limit so we limit it to some bounds. In our case, we have kept a *maximum of 10 and minimum of -10.*
5. We add the double of our acceleration value to the `y` coordinates.
6. We set our deceleration to 0 to reset it on our next `update()` call.

 You can tweak the acceleration values to test for yourself and get comfortable with the desired speed. To change the speed of acceleration, simply change the values in `dyc`.

And this will make our player jump with joy! Build your game and test it on your device or emulator. You will observe two things here:

- The game is paused when you start it and will only play when you touch the screen. This is because of our if condition in our `GameView.java` file.
- Your player now moves up and down because of the acceleration values we provided in the `PlayerCharacter.java` file.

Now that we have established this, it's time to deal with oncoming obstacles and dodge them! Oh, and also learn about collisions in the process.

Dodging incoming rocks

The very first thing we would need before we proceed with anything is a nice looking piece of rock. Not kidding, really! We need an image of a toony-looking rock, which would fit with the theme of our game. So, let's get a rock for ourselves. For the purpose of this chapter, we will be using this rock:

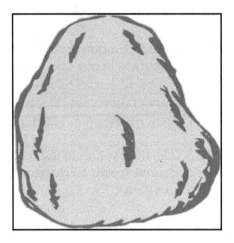

This is our basic rock

As we had done in our previous part, we will need animation for the rock and so we will create a sprite sheet of the rock we have. Then, we will simply name it as `rock.png` and get started with it. It would be cool to have a rolling rock since that would make more sense in the game, and so we will use a rolling rock sprite sheet.

Our **rock.png** sprite sheet

Of course, as we saw before, this rock is also just an image and will be placed inside the res/drawable folder as we did before. So, pick up any rock and place it into the res/drawable folder to get started.

Now, since we are going to deal with a new object it must be pretty clear by this point that we would need to create a new class. So let's go ahead and do that. Create a new class and name it Rock.java. This will have all our code for our obstacle rock. You will then have a blank java file in which you will need to extend it to our GameObj.java file like every other game object in our game:

```
package nikhil.nikmlnkr.game;

/**
 * Created by Nikhil on 30-01-2017.
 */

public class Rock extends GameObj {
}
```

Now gear up! We are going to do some heavy duty coding on this one. Let's take a look at this step by step. We will need three methods in this file:

- Constructor: This will have the parameters for our rock, such as the x, y positions, width, height, score, and so on
- Update method: This is the method, as seen earlier, that gets called in every frame
  - Draw method: This method is responsible for actually drawing our object on the screen

Let's begin with our constructor and variables. We will need a score and speed variable as well as a variable for Bitmap and our predefined AnimationClass, since we will be animating this rock as well. Let's do that. We will start with the variables first. We will declare our variables that will be needed for the game, which includes score, speed, our animationClass reference, and spriteSheet reference. We have also taken a Random number reference variable to generate a score based on a unique condition in the game loop that you will see shortly. Currently, we are not displaying the score on the screen but soon you will see it:

```
package nikhil.nikmlnkr.game;

import android.graphics.Bitmap;
import android.graphics.BitmapFactory;
import android.graphics.Canvas;
```

```
import java.util.Random;

/**
 * Created by Nikhil on 30-01-2017.
 */

public class Rock extends GameObj{

    private int score;
    private int speed;
    private Random rnd = new Random ();
    private AnimationClass animationClass = new AnimationClass();
    private Bitmap spriteSheet;


}
```

Now, we will write our constructor. There are a bunch of parameters needed for the constructor such as the x, y coordinates, `width`, `height`, `score`, and `noOfFrames`. We basically take their reference as parameters to our method and hence we require `xc`, `yc`, `w`, `h`, `s`, and `noOfFrames` as parameters in our constructor.

After that, we actually need our rock to go faster once we survive for a longer time and hence we will write the code in such a way that, as our score increases, our rocks will become faster. We will use a little bit of math for this, wherein we will use our random variable. Basically, this will be our formula: `speed = 7 + (int)` `(rnd.nextDouble()*score/30);`. Then we will set our Bitmap reference variable so it can scan through our sprite sheet followed by a `for` loop to scan through the same. And finally, we will set our frames to our `animationClass` and add a delay for our animation to play. Let's see how that works out in our code:

```
    public Rock (Bitmap res, int xc, int yc, int w, int h, int s, int
noOfFrames) {
        this.xc = xc;
        this.yc = yc;
        width = w;
        height = h;
        score = s;

        speed = 7  + (int) (rnd.nextDouble()*score/30);

        if(speed > 35)
            speed = 35;

        Bitmap[] img = new Bitmap[noOfFrames];

        spriteSheet = res;
```

```
for(int i=0; i<img.length; i++) {
    img[i] = Bitmap.createBitmap(spriteSheet, 0, i*height,
        width,   height);
}

animationClass.setFrames(img);
animationClass.setDelay(100);

}
```

In this part of the code, we did nothing new but simply created our rock game object and its drawing on the screen method. It's the same thing that we did with our `PlayerCharacter`.

Now that our constructor is ready, we can move on to write our code for the `update()` and `draw()` methods. In our `update()` method, we don't need to do much. We'll only be moving our rock from right to left, and hence we'll use our `speed` variable to just shift our rock certain units in every frame toward the left. We will use the `draw()` method to simply draw our rock on the screen with the help of the `animationClass` reference variable:

```
public void update() {
    xc -= speed;
    animationClass.update();
}

public void draw(Canvas canvas) {
    try {
        canvas.drawBitmap(animationClass.getImage(), xc, yc, null);
    } catch (Exception e) {}
}
```

As you can see in the preceding code, we simply created our `update()` method and shifted our rock certain units on the left with our `speed` variable. After that, immediately we gave a call to the `update()` method of `animationClass` so that the sprite sheet of the rock gets updated and we get an effect as though the rock is rolling as it moves.

We then simply used the `drawBitmap()` method to draw our rock on the screen.

We created our object, and now it's time for us to display it on the screen, and once we are done with that objective, we will move on to detecting the collision when it collides with our player.

We will now create our rocks on the `GameView.java` file, which will spawn rocks at continuous intervals. Also, when our player collides with our rocks it will pause the game. So, let's open up our `GameView.java` file and get started with writing the code for this.

We will create three new reference variables, `rockStartTime`, `rocks`, and `rnd`, which correspond to the start time of our rock game object, our actual rock game object, and a random variable for the purpose of randomizing the spawn location of the rocks on the screen. Since we are going to be spawning multiple rocks on the screen, we took its data type to be an `ArrayList` since dynamic arrays can be supported through an `ArrayList`. We will be needing this primarily because after our rocks have left the screen space, we will remove them, thereby having proper memory management of resources:

```
private long rockStartTime;
private ArrayList<Rock> rocks;

private Random rnd = new Random();
```

We will improvise a bit on our `surfaceDestroyed()` method and create a `counter` as well as shift our `retry` variable block in order to avoid an infinite loop situation. Code changes are marked in bold:

```
@Override
public void surfaceDestroyed(SurfaceHolder holder){
    boolean retry = true;
    int counter = 0;
    while(retry && counter <1000)
    {
        counter++;
        try {mainThread.setRunning(false);
            mainThread.join();
            retry = false;
        } catch(InterruptedException e){e.printStackTrace();}
    }

}
```

In our `surfaceCreated()` method, we assign our reference variable rocks with our `Rock` class and we initialize the `rockStartTime` variable to take in the current `System.nanoTime()` as follows:

```
@Override
public void surfaceCreated(SurfaceHolder holder){

    bgImg = new BackgroundImage(BitmapFactory.decodeResource
    (getResources(),  R.drawable.background_image));
    playerCharacter = new
    PlayerCharacter(BitmapFactory.decodeResource
    (getResources(),R.drawable.player_run),200,246,3);
    rocks = new ArrayList<Rock>();
```

```
rockStartTime = System.nanoTime();

//we can safely start the game loop
mainThread.setRunning(true);
mainThread.start();

}
```

In our update method, we address the real beast of *spawning rocks on the screen*. There are multiple things we are managing here so let's break them down further.

We will declare our `rockElapsed` variable, which keeps track of how much time the rock has been on the screen.

Then, we define the frequency with which we want to spawn our rocks on the screen. You can manipulate this value in the `if` condition as you please according to the desired effects you want.

Inside this `if` block, we have another nested `if` block which is essentially just to define where the rock will spawn. Here, we are defining the rock spawning in such a way that the first rock element is in the middle part of the screen, as you can see from the math calculations. Or if it is any other element than our first spawned rock, then we tell it to spawn randomly in any location where our `rnd` random variable comes into the picture.

We then define our collision logic. We will use a `for` loop here to run through all our rock objects on the screen, and if any element collides with our player character, then we pause the game.

Lastly, if the rock goes out of our defined screen space, then we `remove` the object from our `ArrayList`.

Let's write the code for this:

```
public void update(){
    if(playerCharacter.getPlaying()) {
        bgImg.update();
        playerCharacter.update();

        //spawn rocks on screen
        long rockElapsed = (System.nanoTime() -
        rockStartTime/1000000);
        if(rockElapsed>(2000 - playerCharacter.getScore()/4)){
            if(rocks.size() == 0){
                rocks.add(new Rock(BitmapFactory.decodeResource
                (getResources(), R.drawable.rock),
                WIDTH+10, HEIGHT/2, 66, 82,
```

```
                    playerCharacter.getScore(),3));
            } else {
                rocks.add(new Rock(BitmapFactory.decodeResource
                (getResources(), R.drawable.rock),
                WIDTH+10, (int)
                (rnd.nextDouble() * (HEIGHT)), 66, 82,
                playerCharacter.getScore(),3));
            }
            rockStartTime = System.nanoTime();
        }

        for(int i=0; i<rocks.size();i++) {
            rocks.get(i).update();
            if(collision(rocks.get(i),playerCharacter)) {
                rocks.remove(i);
                playerCharacter.setPlaying(false);
                break;
            }
            //remove rocks if they go out of the screen
            if(rocks.get(i).getXC()<-100) {
                rocks.remove(i);
                break;
            }

        }
    }
}
```

We then move on to write our collision detection function. We do this based on the concept of the Bounding Box Collision technique that you learned about earlier in this chapter. You can easily observe from this code, `Rect.intersects(a.getRectangle(),` `b.getRectangle())`, that we are simply comparing the rectangles of both of our objects using the `Rect` class that is predefined in our `android.graphics.Rect` import. If there is an overlapping rectangle, then this function returns a `true` value; otherwise, it returns `false`. This function is of the return type `boolean` so it is important for it to `return` a `boolean` value:

```
public boolean collision(GameObj a, GameObj b) {

        if(Rect.intersects(a.getRectangle(), b.getRectangle())) {
            return true;
        }
        return false;
}
```

And then finally, we draw our rock object on the screen. We do this again with a `for` loop and using our `draw()` method:

```
@Override
public void draw(Canvas canvas)
{
    final float scaleFactorX = getWidth()/WIDTH;
    final float scaleFactorY = getHeight()/HEIGHT;
    if(canvas!=null) {
        final int savedState = canvas.save();
        canvas.scale(scaleFactorX, scaleFactorY);
        bgImg.draw(canvas);
        playerCharacter.draw(canvas);

        for(Rock r : rocks) {
            r.draw(canvas);
        }

        canvas.restoreToCount(savedState);
    }
}
```

This will get our game equipped with our collision detection technique. After writing this code, it must be quite obvious how our collision technique is going to work in our game. We created a function with return type Boolean, which will detect collisions between the rectangles of our two objects, which we pass into it as references.

Let's quickly review our code blocks one by one by listing down exactly what each code block is responsible for in our `GameView.java` file to get a better clarity.

Defining our variables

We define our variables required by adding our desired variables, such as `rockStartTime`, `rocks`, and `rnd`:

```
public static final int WIDTH = 1920;
public static final int HEIGHT = 1080;
public static final int MOVINGSPEED = -5;
private long rockStartTime;
private MainGameThread mainThread;
private BackgroundImage bgImg;
private PlayerCharacter playerCharacter;
private ArrayList<Rock> rocks;

private Random rnd = new Random();
```

These variables give us references of objects to work with.

Tackling the infinite loop situation

In our preceding code, there was a chance of running into an infinite loop and so we replace our retry and add in a counter for extra safety in case an infinite loop situation arises in our `surfaceDestroyed()` method. It is possible that our retry might return a `false` value or a `true` value during every run, and so there's a chance of an infinite loop situation. To avoid this, we have this counter which increments every time the while loop is running and stops after that. You can go ahead and experiment to see what problems occur if you avoid using the counter for yourself:

```
@Override
public void surfaceDestroyed(SurfaceHolder holder){
    boolean retry = true;
    int counter = 0;
    while(retry && counter <1000)
    {
        counter++;
        try{mainThread.setRunning(false);
            mainThread.join();
            retry = false;
        }catch(InterruptedException e){e.printStackTrace();}
    }

}
```

This tackles our infinite loop situation.

Initializing our variables

After defining our variables, it is also essential to initialize them. We do so from our `surfaceCreated()` method as follows:

```
@Override
public void surfaceCreated(SurfaceHolder holder){

    bgImg = new BackgroundImage(BitmapFactory.decodeResource
    (getResources(), R.drawable.background_image));
    playerCharacter = new PlayerCharacter(BitmapFactory.decodeResource
    (getResources(),R.drawable.player_run),200,246,3);
    rocks = new ArrayList<Rock>();

    rockStartTime = System.nanoTime();

    //we can safely start the game loop
```

```
    mainThread.setRunning(true);
    mainThread.start();

}
```

And that takes care of our variable values.

Collision behavior

After this, we define what exactly will happen if there is a collision between two objects, which is defined in our update() method. We also address the situation if the rocks go out of the screen:

```
long rockElapsed = (System.nanoTime() - rockStartTime/1000000);

if(rockElapsed>(2000 - playerCharacter.getScore())){

    if(rocks.size() == 0){
        rocks.add(new Rock(BitmapFactory.decodeResource
        (getResources(), R.drawable.rock),
        WIDTH+10, HEIGHT/2, 66, 82,
        playerCharacter.getScore(),3));
    } else {
        rocks.add(new Rock(BitmapFactory.decodeResource
        (getResources(), R.drawable.rock),
        WIDTH+10, (int) (rnd.nextDouble() * (HEIGHT)), 66, 82,
        playerCharacter.getScore(),3));
    }

    rockStartTime = System.nanoTime();
}

for(int i=0; i<rocks.size();i++) {

    rocks.get(i).update();

    if(collision(rocks.get(i),playerCharacter)) {
        rocks.remove(i);
        playerCharacter.setPlaying(false);
        break;
    }

//remove rocks if they go out of the screen
    if(rocks.get(i).getXC()<-100) {
        rocks.remove(i);
        break;
    }
```

```
    }
```

Our collision behavior is now ready.

Collision function

We defined our collision behavior, but it is also necessary that we write a logic that defines our collision itself. So, we make a collision detection function:

```
public boolean collision(GameObj a, GameObj b) {

    if(Rect.intersects(a.getRectangle(), b.getRectangle())) {
        return true;
    }
    return false;
}
```

Our collision function is now ready.

Drawing our objects on the screen

Once we are done, we simply draw our object on the screen in our if a block of our draw() method:

```
for(Rock r : rocks) {
    r.draw(canvas);
}
```

And that takes care of all the functionality needed to detect a collision.

Now that we are done with our collision logic, let's make a few tweaks to our player character. We have some unused variables there so let's get rid of them and customize it a little further. Prior to this, they were needed to get an understanding how things work. But now they are pretty useless to us, so there's no need for them to be lying around. Open up your PlayerCharacter.java and make the following changes to your code:

```
package nikhil.nikmlnkr.game;

import android.graphics.Bitmap;
import android.graphics.Canvas;

/**
 * Created by Nikhil on 13-01-2017.
 */
```

```
public class PlayerCharacter extends GameObj{
    private Bitmap spriteSheet;
    private int score;
    private boolean up, playing;
    private AnimationClass ac = new AnimationClass();
    private long startTime;

    public PlayerCharacter(Bitmap res, int w, int h, int noOfFrames) {
        xc = 100;
        yc = GameView.HEIGHT/2;
        //removing the dya variable
        dyc = 0;
        score = 0;
        height = h;
        width = w;

        Bitmap[] img = new Bitmap[noOfFrames];
        spriteSheet = res;

        for(int i=0; i<img.length;i++){
            img[i] = Bitmap.createBitmap(spriteSheet, i*width, 0,
            width, height);
        }
        ac.setFrames(img);
        ac.setDelay(10);
        startTime = System.nanoTime();
    }

    public void setUp(boolean b){
        up = b;
    }

    public void update() {
        long elapsed = (System.nanoTime()-startTime)/1000000;
        if(elapsed > 100) {
            score++;
            startTime = System.nanoTime();
        }
        ac.update();

        if(up){
            dyc -=1;
        }
        else {
            dyc +=1;
        }

        if(dyc > 10) {
```

```
                dyc = 10;
        }
        if(dyc < -10) {
            dyc = -10;
        }

        yc += dyc*2;
        //removing the dya variable
    }

    public void draw(Canvas canvas) {
        canvas.drawBitmap(ac.getImage(), xc, yc, null);
    }

    public int getScore() {
        return score;
    }

    public boolean getPlaying(){
        return playing;
    }

    public void setPlaying(boolean b) {
        playing = b;
    }

    public void resetDYC() {
        dyc = 0;
    }

    public void resetScore () {
        score = 0;
    }
}
```

These are just minor tweaks which are quite self-explanatory and don't require any separate explanation as such. Now that we have established all of this, we can finally test our collision technique! So, let's go ahead and do that. Build your game and test it on your device or emulator. You will see something like this:

Hurray! Our rocks are on screen.... But wait, what?

So, the good news is that our rocks finally appear on the screen and the bad news is that they are all over the screen. We will address this situation shortly, but for now, we can get to test our collision functionality. As soon as our rocks collide with our player character the game will pause. We will now need to fix their frequency in order to control their spawning.

Open up your `GameView.java` file and make the changes marked in bold. Remove all commented code marked in bold and add the `if(rocks.size() < 2)` statement as follows:

```java
public class GameView extends SurfaceView implements SurfaceHolder.Callback
{
    public static final int WIDTH = 1920;
    public static final int HEIGHT = 1080;
    public static final int MOVINGSPEED = -5;
    //private long rockStartTime;
    private MainGameThread mainThread;
    private BackgroundImage bgImg;
    private PlayerCharacter playerCharacter;
    private ArrayList<Rock> rocks;

    private Random rnd = new Random();
```

```
//Constructor, surfaceDestroyed and surfceChanged methods remain
same
@Override
public void surfaceCreated(SurfaceHolder holder){

    bgImg = new BackgroundImage(BitmapFactory.decodeResource
    (getResources(), R.drawable.background_image));
    playerCharacter = new
    PlayerCharacter(BitmapFactory.decodeResource
    (getResources(),R.drawable.player_run),200,246,3);
    rocks = new ArrayList<Rock>();

    //rockStartTime = System.nanoTime();

    //we can safely start the game loop
    mainThread.setRunning(true);
    mainThread.start();

}
//onTouchEvent remains same

public void update()
{
    if(playerCharacter.getPlaying()) {
        bgImg.update();
        playerCharacter.update();

        //spawn rocks on screen
        //long rockElapsed = (System.nanoTime() -
        rockStartTime/1000000);
        //if(rockElapsed>(2000 - playerCharacter.getScore())){
         if(rocks.size() < 2){
            if(rocks.size() == 0){
                rocks.add(new Rock(BitmapFactory.decodeResource
                (getResources(), R.drawable.rock),
                WIDTH+10, HEIGHT/2, 120, 82,
                playerCharacter.getScore(),3));
            } else {
                rocks.add(new Rock(BitmapFactory.decodeResource
                (getResources(), R.drawable.rock),
                WIDTH+10, (int) (rnd.nextDouble() * (HEIGHT)),
                120, 82, playerCharacter.getScore(),3));
            }
            rockStartTime = System.nanoTime();
        } //Bracket ends here

        for(int i=0; i<rocks.size();i++) {
            rocks.get(i).update();
```

```
            if(collision(rocks.get(i),playerCharacter)) {
                rocks.remove(i);
                playerCharacter.setPlaying(false);
                break;
            }
            //remove rocks if they go out of the screen
            if(rocks.get(i).getXC()<-100) {
                rocks.remove(i);
                break;
            }

        }
    }
}

public boolean collision(GameObj a, GameObj b) {

    if(Rect.intersects(a.getRectangle(), b.getRectangle())) {
        return true;
    }
    return false;
}

@Override
public void draw(Canvas canvas)
{
    //same as the draw method seen earlier
}

}
```

We simply commented out our `rockElapsed` and `rockStartTime` variables and instead replaced our `if` condition with the `rocks.size()` condition, which tells our game to spawn only 2 rocks on the screen at any given time. This way, we get a control on the frequency of rocks spawned on the screen. You can tweak your frequency as you desire:

Okay, this looks better than the previous one

Of course, there is still the question of the appearance of the rock; this depends on the image that you use to create the sprite animation. But for now, we will move ahead with the further part. Also, note that in a similar way, you can also create collectible items such as coins so it is recommended for you create a coins class, which the player can collect and add to his score. Go ahead, experiment!

Now let's learn about our next concept, that is artificial intelligence. Since we are not going to use any artificial intelligence in our game, we will just go through the concept of this topic. Our current game is a simple side scroller, which does not require any enemies as such. So, let's get started with artificial intelligence, or as many call it, AI And of course, if you come up with an idea to add enemies in the game then you can always use the concepts you learned from the previous chapters and couple them with the concept you are about to learn to come up with something cool!

Artificial Intelligence

Artificial Intelligence (A.I.) is the study of systems being able to perform human tasks and automation without the requirement of human involvement. In games, this concept is widely used in order to create a real behavior for game enemies. Let's understand the basic concept of A.I. in this chapter. This part of the chapter is going to be theory only so feel free to skip this part if you just want to focus on getting your game ready. However, it is highly recommended to go through this part because if you want to create A.I. in your game then the concepts from this part will come in handy.

History of Artificial Intelligence

In games, A.I. is used to create intelligent behaviors to objects that the user cannot control. This can be anything ranging from a dragon you see on the screen or simply a character that keeps following you. To put it simply, it is something that provides the means for a seemingly inanimate object to possess human-like intelligence. On a technical level, A.I. is the study of coming up with algorithms which include techniques from robotics, computer graphics, computer science, and control theory. There are many algorithms for creating a life-like A.I.

The study of A.I. in games has been a part of this industry ever since its inceptions, as games provide one of the best ways to simulate artificial behavior in lifeless objects. However, if you take a look back at the 1950s era, you will observe that A.I. was relatively a simpler concept than what it has evolved into in today's times. The game of Nim made in 1951 was the first examples of A.I.:

Game of Nim in 1951

This game was a simple Tic Tac Toe game, which demonstrated the very first example of A.I. while playing against the computer. The first notable ones for the arcade appeared in 1974: the Taito game, Speed Race (a racing video game), and the Atari games, Qwak (a duck hunting light gun shooter) and Pursuit (a fighter aircraft dogfighting simulator). Two text-based computer games from 1972, Hunt the Wumpus and Star Trek, also had enemies. Enemy movement was based on stored patterns. The incorporation of microprocessors would allow more computation and random elements overlaid into movement patterns. Let's talk about some interesting A.I. algorithms.

Artificial Intelligence Algorithms

Algorithms for A.I. can include many aspects. Essentially, to break these down into simple terms, they are just search algorithms. Following are some of the most popular search algorithms, which can be implemented into A.I. logic. In between, there are pseudo codes for understanding these examples. Once you go through these, we can come up with some basic A.I. logic for our game if you wish to include it from your side.

Breadth-First Search

This algorithm starts from the root node and then, after exploring all the neighboring nodes first, it moves toward the next level of neighbors, which returns the shortest path to our solution. It uses the FIFO queue data structure.

The disadvantage of this algorithm is that it consumes a lot of memory because each level of nodes is saved in order to create the next one.

Read more at: `https://en.wikipedia.org/wiki/Breadth-first_search`.

Depth-First Search

The implementation of this algorithm is done using recursion with the LIFO stack data structure and it creates the same nodes as our first method but in a different order. The space requirement is quite linear since nodes on a single path are stored in each iteration from root-to-leaf node.

The disadvantage of the algorithm is that there's a possibility that this algorithm may not terminate and go on infinitely on one path and in some cases the execution time increases. It can't check for duplicate nodes.

Read more at: `https://en.wikipedia.org/wiki/Depth-first_search`

Bidirectional search

In this technique, the search is carried forward from the initial state and backward from the goal state until both are met to identify the common state and then the path from the initial state is concatenated with an inverse path from the goal state.

Read more at: `https://en.wikipedia.org/wiki/Bidirectional_search`

Uniform cost search

In this algorithm, the sorting is done by increasing the cost of the path to a node and the node with the least cost is expanded. It is also popularly known as Dijkstra's algorithm.

The disadvantage of this algorithm is that, since there can be multiple long paths, this technique must explore them all.

Read more at: https://en.wikipedia.org/wiki/Dijkstra%27s_algorithm

Iterative deepening Depth-First Search

A depth-first search to level 1 is performed and then the same is done on level 2 and so on, until the solution is found. Until all the lower nodes are generated, a node is not created.

Read more at: https://en.wikipedia.org/wiki/Iterative_deepening_depth-first_se arch

Comparison of preceding algorithms complexities

Let's take a look at some of the interesting results by comparing our first five algorithms. Here's the performance of the algorithms based on various criteria:

Criteria	Breadth First	Depth First	Bidirectional	Uniform Cost	Deepening
Time	b^d	b^m	$b^{d/2}$	b^d	b^d
Space	b^d	b^m	$b^{d/2}$	b^d	b^d
Optimality	Yes	No	Yes	Yes	Yes
Completeness	Yes	No	Yes	Yes	Yes

A * search

This algorithm is the best known as the Best First search algorithm and is also widely used in games for pathfinding. Its performance is very efficient and expanding paths that are already expensive are avoided through this algorithm.

$f(n) = g(n) + h(n)$, where $g(n)$ the cost (so far) to reach the destination node, $h(n)$ is the estimated cost to get from the node to the goal, and $f(n)$ is the estimated total cost of the path through n to the goal. It is implemented using a priority queue by increasing $f(n)$.

Read more at: `https://en.wikipedia.org/wiki/A*_search_algorithm`

Creating your own Artificial Intelligence

Now, as for this game, we don't really require any A.I. since most of our obstacles themselves are a big challenge for us. However, we can have some food for thought. Here's an exercise for you. Using your knowledge gained in previous chapters, try to create an A.I. class for an enemy spawning, which will start shooting projectiles at you if it enters a certain radius. Here are your objectives for creating an A.I.:

- Spawn enemies from right side of the screen
- Make them move in the left-hand direction
- Upon entering a certain radius, they will start shooting projectiles at you
- If you collide with the projectile, the game pauses

You can use your Rock.java class as a shooting projectile. You can have a monkey coming in from the side of the screen which starts throwing rocks at you if it enters a certain radius. Try this out by yourself; it'll be fun to see what you come up with!

So, that's about it for our collision and A.I. part of this book.

Summary

In this chapter, you learned a great deal about the various collision techniques as well as multiple artificial intelligence algorithms. We now know how to create collisions based on the Bounding Box Collision technique.

In the next chapter, we will add a ground for our player and understand how we can create explosions based on our collisions.

7
Adding Boundaries and Using Sprites to Create Explosions

In our preceding chapter, we covered the part where we successfully detected collisions. Now that we have understood how to deal with collisions, we can play around with our knowledge and add some cool stuff to our game. This chapter will be pretty concise, and if you have understood the concept of collision detection properly, then this will be a breeze for you. Here's what we will do in this chapter:

- Add a ground for our player as a boundary
- Detect a collision between our player and rocks
- Spawn an explosion sprite on the point where a collision takes place

Observe that from the preceding tasks, we have already accomplished our second task, so we have to focus on the first and third tasks. Let's dive into creating a ground for our player because right now our player is simply going down infinitely. We will also add an upper boundary in order to contain our player within the screen, otherwise, our player would go right outside the screen. Again, we will divide this chapter into two parts:

- Adding boundaries to our game
- Using sprites to create explosions

Let's get started!

Adding boundaries

As we are quite familiar with the process of creating a new class, we will simply create our two new classes for our upper and lower boundary and call them `UpperBoundary.java` and `LowerBoundary.java`, respectively. We have the following objectives for our boundaries:

- Make them appear at the top and bottom of our game screen
- If a player collides with them, then reset the game

With these objectives in mind, we will move ahead to create boundaries for our game.

Creating the classes for our boundaries

Before actually creating our boundaries, we will need an image sprite in order to make them visible on the screen. For this purpose, we will take a simple sprite with a plain color. Here's the sprite that we will be using for our game:

Our ground.png file

Also, we will place this file in our `res/drawable` folder as we did for our previous image files. Once you are done with this, move on to the next part.

So now, let's create our `UpperBoundary.java` class. Go ahead, create a new class and write the following code in it:

```
public class UpperBoundary extends GameObj {
    private Bitmap img;

    public UpperBoundary(Bitmap res, int xc, int yc, int h) {
        height = h;
        width = 20;

        this.xc = xc;
        this.yc = yc;

        dxc = GameView.MOVINGSPEED;
        img = Bitmap.createBitmap(res, 0, 0, width, height);
    }

    public void update(){
        xc += dxc;
    }

    public void draw(Canvas canvas) {
        try{
            canvas.drawBitmap(img, xc, yc, null);
        } catch(Exception e) {
        };
    }
}
```

In this code, we are simply creating our boundary class with its constructor and `update()` and `draw()` functions. The constructor is pretty simple to understand here. Whenever we create an instance of our boundaries, we will pass in a sprite, the x y position, and the height. We will also set a speed in which the boundaries will move backward, giving an illusion that our player is moving forward.

Like all our previous game objects, this class also extends our main `GameObj` class. In a similar way, we will also create our `LowerBoundary.java` class. The only change we will make is in our `height` and `width` variable and the rest of the entire code remains the same as that of the `UpperBoundary.java` file:

```
public class LowerBoundary extends GameObj {
    public LowerBoundary(Bitmap res, int xc, int yc) {
        height = 200;
        width = 20;
    }
}
```

Note here that we are keeping the `height` constant since we want to spawn our bottom border as low as possible and thereby `200` is a safe limit for the same. You can experiment with these values as per your liking.

Now that we have our classes ready, it is time for us to bring them into our game.

Creating boundaries in our game

We will go step by step here since this part may get a bit tricky. We will have to deal with a lot of math calculations for our boundaries as well as lots and lots of new variables. Let's start by declaring some variables in our `GameView.java` file.

Throughout this part, we will be working only in our `GameView.java` file.

Creating the variables required

Here are the new variables that we will declare:

```
private ArrayList<UpperBoundary> upperBoundary;
private ArrayList<LowerBoundary> lowerBoundary;

private int maxBoundaryHeight;
private int minBoundaryHeight;

private boolean upBound = true;
private boolean lowBound = true;

private int progressDenom = 20;

private boolean newGameCreated;
```

We create our `upperBoundary` and `lowerBoundary` variables as `ArrayList` to keep a track of our actual game object on the screen, then we also create two integer variables--`maxBoundaryHeight` and `minBoundaryHeight`--to keep a track of the maximum and minimum heights for our upper boundary. We also create two Boolean variables--`upBound` and `lowBound`--if our boundaries go out of our specified minimum or maximum height. The `progressDenom` variable is created as an integer in order to create a cool pattern for our ground rather than just being plane. Finally, we have a `newGameCreated` Boolean variable, which will reset our game automatically if our player crashes with any object.

We are set with our variables. We will now move on to referencing our boundaries when the game is started.

Referencing our boundaries

Just as we assigned a value to our reference in our rocks, we will do the same for both our boundaries. We will do so in our `surfaceCreated()` method by adding the variables marked in bold:

```
    @Override
    public void surfaceCreated(SurfaceHolder holder){
upperBoundary = new ArrayList<UpperBoundary>();
    lowerBoundary = new ArrayList<LowerBoundary>();
    }
```

Looks neat! Now comes the tricky part. We will have to write the update logic for both our boundaries. Don't confuse this with the `update()` method in our individual boundaries. That update method will simply make our ground move backward. We will also need to actually write the logic to spawn them on screen. Let's see how to do that.

Updating our boundaries

We will be updating our upper boundaries on every 50th score and lower boundaries on every 40th score. Let's write the code for our boundaries. This involves a lot of tricky mathematical calculations, so watch out. However, contradictory to that, every step is quite self-explanatory here. Here's the base logic for our boundaries:

1. Update on every 50th or 40th score
2. Add our image onto the screen
3. After every frame, call the `update()` method in our boundary class
4. If the boundary goes out of screen, then remove it
5. If either of the boundary exceeds its maximum or minimum value, then accordingly set its `upBound` or `lowBound` variable to either `true` or `false`, depending on its position

This is the logic for our boundaries, and the same is repeated for both our upper and lower boundaries. This code block is written after our `draw()` method. We write the code for them as follows:

```
public void updateUpperBound () {
    if(playerCharacter.getScore() % 50 == 0){
        upperBoundary.add(new
        UpperBoundary(BitmapFactory.decodeResource
        (getResources(), R.drawable.ground),
        upperBoundary.get(upperBoundary.size()-1).
        getXC() + 20, 0, (int)((rnd.nextDouble()*
```

```
            (maxBoundaryHeight))+1)));
    }

    for(int i=0; i<upperBoundary.size();i++) {
        upperBoundary.get(i).update();
        if(upperBoundary.get(i).getXC() < -20){
            upperBoundary.remove(i);


            if(upperBoundary.get(upperBoundary.size()-1).
            getHeight() >= maxBoundaryHeight) {
                upBound = false;
            }


            if(upperBoundary.get(upperBoundary.size()-1).
            getHeight() <= minBoundaryHeight) {
                upBound = true;
            }

            if(upBound){
                upperBoundary.add(new
                UpperBoundary(BitmapFactory.decodeResource
                (getResources(), R.drawable.ground),
                upperBoundary.get(upperBoundary.size()-1).
                getXC() + 20, 0,upperBoundary.get
                (upperBoundary.size()-1).getHeight()+1));
            } else {
                upperBoundary.add(new
                UpperBoundary(BitmapFactory.decodeResource
                (getResources(), R.drawable.ground),
                upperBoundary.get(upperBoundary.size()-1).
                getXC() + 20, 0, upperBoundary.get
                (upperBoundary.size()-1).getHeight()-1));
            }
        }
    }
}

public void updateLowerBound () {
    if(playerCharacter.getScore() % 40 == 0) {
        lowerBoundary.add(new
        LowerBoundary(BitmapFactory.decodeResource
        (getResources(),R.drawable.ground),
        lowerBoundary.get(lowerBoundary.size() - 1).
        getXC() + 20,(int)((rnd.nextDouble()*
        maxBoundaryHeight) + (HEIGHT - maxBoundaryHeight))));
    }
```

```
for(int i=0;i<lowerBoundary.size();i++) {
    lowerBoundary.get(i).update();

    if(lowerBoundary.get(i).getXC()<-20){
        lowerBoundary.remove(i);

        if(lowerBoundary.get(lowerBoundary.size()-1).
        getHeight() >= maxBoundaryHeight) {
            lowBound = false;
        }


        if(lowerBoundary.get(lowerBoundary.size()-1).
        getHeight() <= minBoundaryHeight) {
            lowBound = true;
        }

        if(lowBound) {
            lowerBoundary.add(new
            LowerBoundary(BitmapFactory.decodeResource
            (getResources(),R.drawable.ground),
            lowerBoundary.get(lowerBoundary.size() - 1).
            getXC() + 20, lowerBoundary.
            get(lowerBoundary.size() - 1).getYC() + 1));
        } else {
            lowerBoundary.add(new
            LowerBoundary(BitmapFactory.decodeResource
            (getResources(),R.drawable.ground),
            lowerBoundary.get(lowerBoundary.size() - 1).
            getXC() + 20, lowerBoundary.
            get(lowerBoundary.size() - 1).getYC() - 1));
        }
    }
}
}
```

Now, we have to actually draw our boundaries on our screen, so we will go to our draw() method to do so.

Drawing our boundaries on the screen

As with our previous images, we use the `draw()` method to write our code for displaying our ground on the screen:

```
for(UpperBoundary ub : upperBoundary){
    ub.draw(canvas);
}

for(LowerBoundary lb: lowerBoundary) {
    lb.draw(canvas);
}
```

Even this part is taken care of for now. Now, we have to look at the collision part of our ground. We need to detect a collision on our ground with the player.

Detecting a collision between the ground and player

Since we have already created our collision method, we simply go ahead and use the function. We already have a clear understanding of how a collision works because of the previous chapter, so we write the following code in our `update()` method of our `GameView.java` file:

```
for(int i=0; i<lowerBoundary.size();i++) {
    if(collision(lowerBoundary.get(i),playerCharacter)) {
        playerCharacter.setPlaying(false);
    }
}

for(int i=0; i<upperBoundary.size();i++) {
    if(collision(upperBoundary.get(i),playerCharacter)) {
        playerCharacter.setPlaying(false);
    }
}
```

Collision detection between the player and the ground is completed. Now, we also have to assign our maximum and minimum boundary heights as well as tweak them as per our `progressDenom`.

Maximum and minimum boundary heights

In our `update()` method, we will assign these values based on the player score and `progressDenom`. We will also use this method to call our `updateUpperBound()` and `updateLowerBound()` methods that we created earlier in this chapter:

```
this.updateUpperBound();
this.updateLowerBound();

maxBoundaryHeight = 30+playerCharacter.getScore() / progressDenom;

if(maxBoundaryHeight > HEIGHT/4)maxBoundaryHeight = HEIGHT/4;
minBoundaryHeight = 5 + playerCharacter.getScore()/progressDenom;
```

We're almost there. Now, the only thing remaining to create is our `newGame()` function.

Creating a new game

We will create a `newGame()` function that will get called every time the player crashes with an object. We're doing nothing, but just resetting our objects as they were initially when we started the game. So, here's how we do it:

```
public void newGame () {
    lowerBoundary.clear();
    upperBoundary.clear();
    rocks.clear();

    minBoundaryHeight = 5;
    maxBoundaryHeight = 30;

    playerCharacter.resetScore();
    playerCharacter.resetDYC();
    playerCharacter.setYC(HEIGHT/2);

    for(int i = 0; i * 20 < WIDTH + 40;i++) {
        if(i == 0) {
            upperBoundary.add(new
            UpperBoundary(BitmapFactory.decodeResource
            (getResources(),R.drawable.ground),
            i * 20, 0, 10));
        } else {
            upperBoundary.add(new
            UpperBoundary(BitmapFactory.decodeResource
            (getResources(),R.drawable.ground),
            i * 20, 0, upperBoundary.get(i - 1).getHeight() + 1));
        }
```

```
    }

    for(int i = 0; i*20<WIDTH+40;i++) {
        if(i==0) {
            lowerBoundary.add(new
            LowerBoundary(BitmapFactory.decodeResource
            (getResources(),R.drawable.ground),
            i * 20, HEIGHT - minBoundaryHeight));
        } else {
            lowerBoundary.add(new
            LowerBoundary(BitmapFactory.decodeResource
            (getResources(),R.drawable.ground),
            i * 20, lowerBoundary.get(i - 1).getYC() - 1));
        }
    }
    newGameCreated = true;
}
```

Also, we still have to call this function from somewhere. As per our objective, we need it to be called after our player crashes. So, we add an else block in our update function after our `if(playerCharacter.getPlaying())` condition, as follows:

```
else {
    newGameCreated = false;
    if(!newGameCreated) {
        newGame();
    }
}
```

We're ready with our code. Let's review our code for changes marked in bold and check whether you missed any step:

```
package nikhil.nikmlnkr.game;

import android.content.Context;
import android.graphics.Bitmap;
import android.graphics.BitmapFactory;
import android.graphics.Canvas;
import android.graphics.Rect;
import android.view.MotionEvent;
import android.view.SurfaceHolder;
import android.view.SurfaceView;

import java.util.ArrayList;
import java.util.Random;


public class GameView extends SurfaceView implements SurfaceHolder.Callback
```

```
{
    public static final int WIDTH = 1920;
    public static final int HEIGHT = 1080;
    public static final int MOVINGSPEED = -5;
    private MainGameThread mainThread;
    private BackgroundImage bgImg;
    private PlayerCharacter playerCharacter;
    private ArrayList<Rock> rocks;
    //Our new variable names
    private ArrayList<UpperBoundary> upperBoundary;
    private ArrayList<LowerBoundary> lowerBoundary;

    private int maxBoundaryHeight;
    private int minBoundaryHeight;

    private boolean upBound = true;
    private boolean lowBound = true;

    private int progressDenom = 20;

    private boolean newGameCreated;

    private Random rnd = new Random();

    public GameView(Context context){
        super(context);
        //set callback to the surfaceholder to track events
        getHolder().addCallback(this);

        mainThread = new MainGameThread(getHolder(), this);

        //make gamePanel focusable so it can handle events
        setFocusable(true);
    }

    @Override
    public void surfaceChanged(SurfaceHolder holder, int format, int width,
int height){}

    @Override
    public void surfaceDestroyed(SurfaceHolder holder){
        boolean retry = true;
        int counter = 0;
        while(retry && counter <1000){
            counter++;
            try{
                mainThread.setRunning(false);
                mainThread.join();
```

```
                retry = false;
            }catch(InterruptedException e){e.printStackTrace();
            }
        }
    }

    @Override
    public void surfaceCreated(SurfaceHolder holder){

        bgImg = new
        BackgroundImage(BitmapFactory.decodeResource
        (getResources(), R.drawable.background_image));

        playerCharacter = new
        PlayerCharacter(BitmapFactory.decodeResource
        (getResources(),R.drawable.player_run),200,246,3);

        rocks = new ArrayList<Rock>();
        //Referencing our upperBoundary and lowerBoundary variables
        upperBoundary = new ArrayList<UpperBoundary>();
        lowerBoundary = new ArrayList<LowerBoundary>();

        //we can safely start the game loop
        mainThread.setRunning(true);
        mainThread.start();
    }
    @Override
    public boolean onTouchEvent(MotionEvent event){
        if(event.getAction() == MotionEvent.ACTION_DOWN) {
            if(!playerCharacter.getPlaying()){
                playerCharacter.setPlaying(true);
                playerCharacter.setUp(true); //minor change
            } else {
                playerCharacter.setUp(true);
            }
            return true;
        }

        if(event.getAction() == MotionEvent.ACTION_UP){
            playerCharacter.setUp(false);
            return true;
        }

        return super.onTouchEvent(event);
    }

    public void update()
    {
```

```
if(playerCharacter.getPlaying()) {
    bgImg.update();
    playerCharacter.update();

    this.updateUpperBound();
    this.updateLowerBound();

    maxBoundaryHeight =
    30 + playerCharacter.getScore() / progressDenom;

    if(maxBoundaryHeight > HEIGHT/4)
        maxBoundaryHeight = HEIGHT/4;

    minBoundaryHeight =
    5 + playerCharacter.getScore()/progressDenom;
    for(int i=0; i<lowerBoundary.size();i++) {
        if(collision(lowerBoundary.get(i),
        playerCharacter)) {
            playerCharacter.setPlaying(false);
        }
    }

    for(int i=0; i<upperBoundary.size();i++) {
        if(collision(upperBoundary.get(i),
        playerCharacter)) {
            playerCharacter.setPlaying(false);
        }
    }

    //spawn rocks on screen
    if(rocks.size() < 2){
        if(rocks.size() == 0){
            rocks.add(new
            Rock(BitmapFactory.decodeResource
            (getResources(), R.drawable.rock),
            WIDTH + 10, HEIGHT/2, 200, 200,
            playerCharacter.getScore(),3));
        } else {
            rocks.add(new
            Rock(BitmapFactory.decodeResource
            (getResources(), R.drawable.rock),
            WIDTH+10, (int) (rnd.nextDouble() *
            (HEIGHT - maxBoundaryHeight * 2))
            + maxBoundaryHeight, 200, 200,
            playerCharacter.getScore(),3));
        }
    }
```

```
            for(int i=0; i<rocks.size();i++) {
                rocks.get(i).update();
                if(collision(rocks.get(i),playerCharacter)) {
                    rocks.remove(i);
                    playerCharacter.setPlaying(false);
                    break;
                }
                //remove rocks if they go out of the screen
                if(rocks.get(i).getXC() < -100) {
                    rocks.remove(i);
                    break;
                }

            }
        } else {
            //We created an else block to trigger our newGameCreated
            variable and to set a new game
            newGameCreated = false;
            if(!newGameCreated) {
                newGame();
            }
        }
    }

    public boolean collision(GameObj a, GameObj b) {
        if(Rect.intersects(a.getRectangle(), b.getRectangle())) {
            return true;
        }
        return false;
    }

    @Override
    public void draw(Canvas canvas){
        final float scaleFactorX = getWidth()/WIDTH;
        final float scaleFactorY = getHeight()/HEIGHT;
        if(canvas != null) {
            final int savedState = canvas.save();
            canvas.scale(scaleFactorX, scaleFactorY);
            bgImg.draw(canvas);
            playerCharacter.draw(canvas);

            for(Rock r : rocks) {
                r.draw(canvas);
            }
            //Drawing our upperBoundary
            for(UpperBoundary ub : upperBoundary){
                ub.draw(canvas);
            }
```

```
            //Drawing our lowerBoundary
            for(LowerBoundary lb: lowerBoundary) {
                lb.draw(canvas);
            }

            canvas.restoreToCount(savedState);
        }
    }

    public void updateUpperBound () {
        //Refer code block above
    }

    public void updateLowerBound () {
        //Refer code block above for this
    }

    public void newGame () {
        //Refer code block in above part
    }
```

Now, build and run your game; we have our boundaries ready!:

Our player won't fall down infinitely now

We can now proceed to our next part of creating explosions on our screen.

Creating explosions

We're almost done with our game here, and only the following two parts are remaining:

- Adding particle effects of an explosion
- Displaying our score on the screen

We will divide this part into two sections wherein we will finish half of our explosions in this chapter, and the further half will be completed along with the UI of our game that will then conclude this game. So, let's get started with this now. For our explosion, we will be needing a sprite sheet. We will use the following sprite sheet for our game:

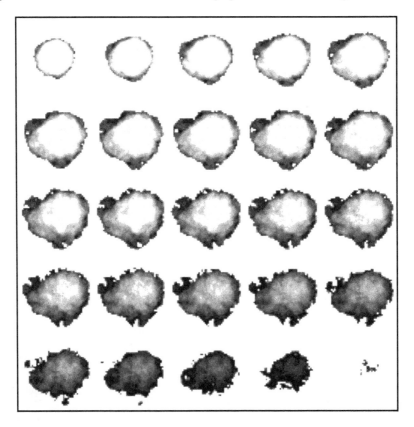

Our explosion sprite sheet

We will create a new class named `ExplosionEffect.java`. Note here that we will not be extending this file to our `GameObj` file since we don't need any of the collision components of this image. We will simply spawn it on the screen and make it stay at the same place. So, open up your `ExplosionEffect.java` file and let's define our variables first:

```
private int xc;
private int yc;
private int height;
private int width;
private int row;
private AnimationClass ac = new AnimationClass();
private Bitmap spriteSheet;
```

As you can see, we just need the x, y coordinates and the `height` and `width` as integer values. Also, observe here that we will be working with both rows and columns in this sprite sheet as opposed to our previous sprites where we just used a singular row or column and hence we will need an extra variable `row` to help us out with this problem. We need our `AnimationClass` variable in order to run our animation and, last but not the least, our `Bitmap spriteSheet` variable.

After this, we will define the constructor for our class as follows:

```
public ExplosionEffect(Bitmap res, int xc, int yc, int w, int h, int
noOfFrames){
    this.xc = xc;
    this.yc = yc;
    this.width = w;
    this.height = h;

    Bitmap[] img = new Bitmap[noOfFrames];

    spriteSheet = res;

    for(int i = 0; i < img.length; i++) {
        if(i % 5 == 0 && i > 0)
            row++;
        img[i] = Bitmap.createBitmap
        (spriteSheet, (i - (5 * row)) * width, row * height,
        width, height);
    }
    ac.setFrames(img);
    ac.setDelay(10);
}
```

If you observe this closely, you will see that we are simply repeating our steps that we did for our previous game objects, except that we have an extra `row` variable here that we defined earlier, and this will help us scan through the rows of our sprite sheet.

Now, we are left with the `draw()` and `update()` methods for this class. We will also make a method to `getHeight()` of the sprite sheet in order for us to work on our calculations when we use this to actually spawn our explosion effects:

```
public void draw(Canvas canvas) {
    if(!ac.playedOnce()){
        canvas.drawBitmap(ac.getImage(),xc,yc,null);
    }
}

public void update() {
    if(!ac.playedOnce()){
        ac.update();
    }
}

public int getHeight() {
    return height;
}
```

Once you are done with these, ensure that your `ExplosionEffect.java` file looks like this:

```
public class ExplosionEffect {

    //refer variables created above
    public ExplosionEffect(Bitmap res, int xc, int yc, int w, int h, int
noOfFrames){
        this.xc = xc;
        this.yc = yc;
        this.width = w;
        this.height = h;

        Bitmap[] img = new Bitmap[noOfFrames];

        spriteSheet = res;

        for(int i = 0; i < img.length; i++) {
            if(i % 5 == 0 && i > 0)
                row++;
            img[i] = Bitmap.createBitmap
            (spriteSheet, (i - (5 * row)) * width, row * height,
width, height);
```

```
        }
        ac.setFrames(img);
        ac.setDelay(10);
    }

    public void draw(Canvas canvas) {
        if(!ac.playedOnce()){
            canvas.drawBitmap(ac.getImage(),xc,yc,null);
        }
    }

    public void update() {
        if(!ac.playedOnce()){
            ac.update();
        }
    }

    public int getHeight() {
        return height;
    }
}
```

We are all set with our `ExplosionEffect.java` file; that's it for this chapter. We are ready with our class for our explosion, and we will start creating explosion effects on our screen after our rocks collide with our player in the next chapter.

Summary

We learned how to create boundaries for our game and also created our explosion effect file with the required components needed along with its constructor.

We now have a proper upper and lower boundary in our game and we also have created our foundation for adding explosions in our game after our player collides with rocks.

In the next chapter, we will spawn the explosion on the screen after our player collides with the rock and display our score on the screen as a User Interface component.

8

Adding an Explosion and Creating a UI

Congratulations on having made it so far! By now, you must be equipped with almost all of the basics needed for you to get started on your journey of game development. This chapter will serve as a finishing touch for our game, and we will complete our explosion part by adding an explosion to our game scene. Once we are done with that, we will create a simple UI for our game that will include displaying our score and distance on the screen. So, buckle up! You are about to finish the game we started. We will be learning the following in this chapter:

- Adding an explosion to our game
- Creating a tutorial with instructions for the player
- Displaying the score on the screen with our UI

However, before we move on to our UI, let's first finish up with our explosion effect.

Adding an explosion to our game

In *Chapter 7*, *Adding Boundaries and Using Sprites to Create Explosions*, we have already created our `ExplosionEffect.java` class file. Now, we are left with just one task: to spawn our explosion on the screen. Now, just for reference, we will take a look at the image we will use for our explosion file:

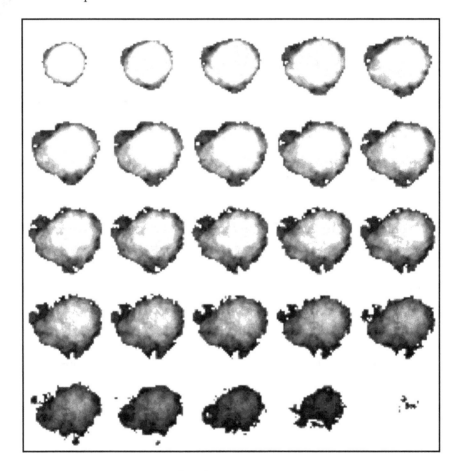

`Explosion.png` file sprite sheet

Note that the last frame of our explosion has almost no image in it. This is because we are not going to destroy this object; we are simply going to spawn it and let it play its animation for now.

Let's proceed and get our explosion running on our game screen. Again, here we will break down our process into simple steps.

Creating variables

As you should be aware by now, we will work on our `GameView.java` file to display our explosion. So, open up your `GameView.java` file. We will start by creating a few variables, as follows:

```
private ExplosionEffect explosionEffect;
private long startReset;
private boolean reset;
private boolean started;
```

We created the `explosionEffect` variable to get a reference to our `ExplosionEffect` class and the other variables we have created in order to make our player reset after it collides. So, basically, what we will do is that once the player collides with a rock, an explosion animation will play and the player will get reset to the initial state.

Some optimizations and improvements

We will do some optimizations to our game by shifting our `mainThread` after our surface has been created. So, we will remove `mainThread = new MainGameThread(getHolder(), this);` from our constructor and write it just above the place where we start running it in our `surfaceCreated()` method, as follows:

```
@Override
public void surfaceCreated(SurfaceHolder holder){

    bgImg = new BackgroundImage(BitmapFactory.decodeResource
    (getResources(), R.drawable.background_image));
    playerCharacter = new PlayerCharacter(BitmapFactory.decodeResource
    (getResources(),R.drawable.player_run),200,246,3);
    rocks = new ArrayList<Rock>();
    upperBoundary = new ArrayList<UpperBoundary>();
    lowerBoundary = new ArrayList<LowerBoundary>();

    mainThread = new MainGameThread(getHolder(), this);
    //we can safely start the game loop
    mainThread.setRunning(true);
    mainThread.start();

}
```

Looks neat! Now, we will also need to make a few improvements to our touch events because we will be resetting our game to its initial state if our player collides with a rock. So, we need to make sure that we can play only if our game is in a play mode or if a new game has been created or reset. So, we will modify our onTouchEvent() to look somewhat like this:

```
@Override
public boolean onTouchEvent(MotionEvent event)
{
    if(event.getAction() == MotionEvent.ACTION_DOWN) {
        if(!playerCharacter.getPlaying() && newGameCreated && reset){
            playerCharacter.setPlaying(true);
            playerCharacter.setUp(true);
        }
        if(playerCharacter.getPlaying()){
            if(!started) started = true;
            reset = false;
            playerCharacter.setUp(true);
        }
        return true;
    }
}
```

Here, we are simply setting up a few parameters for getting more control over our player movement. As you can see from our first if block, if getPlaying(), newGameCreated, and reset return a false value, then we will setPlaying to true and setUp to true.

Also, if our getPlaying() is already true, then we will check whether the game has started; if not, we will set our started variable to true, reset to false, and setUp() to true.

Now, we are set with our new game, playing, and reset logic for playing the game. We handle our reset variables here, but we also have to handle them in the newGame() function that we created earlier. However, before we do that, let's reference our explosion effect and tell it exactly where and when to spawn.

Spawning our explosion

We need our explosion to spawn after our player has collided with our rocks. Once our player collides with a rock, the game is over. So, all of that is handled in our `update()` function wherein we have already created an else block for our `newGame()` function to be called. Let's utilize that and write our logic to spawn an explosion. Our objective here is as follows:

- Spawn the explosion effect
- Start a counter to wait for a certain time after a collision
- Reset the game after a certain time

Note here that the dimensions of our image are 500 x 500, so we will be splitting our image into equal parts for each of our frames, thereby giving us 25 parts, which are 100 x 100 in dimension. We will pass each into our constructor as width, height, and number of frames. If the dimensions of the image that you are using for your game are different, then you need to calculate your dimensions and then use the values as per your image size.

We will go to our `else` block of the `update()` function and modify our previously written code to look like this:

```
else {
    playerCharacter.resetDYC();
    if(!reset) {
        newGameCreated = false;
        startReset = System.nanoTime();
        reset = true;
        explosionEffect = new ExplosionEffect(BitmapFactory
        .decodeResource(getResources(),R.drawable.explosion)
        playerCharacter.getXC(),playerCharacter.getYC()
        -30,100,100,25);
    }

    //Code block after this part remains the same
}
```

So, here we are resetting our player's `y` acceleration and spawning our explosion. Then, after waiting for a certain period of time, we call in the function to reset our game.

Drawing the explosion on the screen

We still have to draw our explosion effect on the screen, and yes, you are right! We will do so in our `draw()` method. We also have to make sure that we draw our explosion only once, that is, at the start of the game, and so we will use our started variable to keep track of it:

```
if(started) {
    explosionEffect.draw(canvas);
}
```

We're all set with our draw logic as well. We're done with the entire logic for our explosion effect, and now your entire code for the `GameView.java` file should look somewhat like the following; all the changes done in this chapter are marked in bold:

```
//Package name and import statements remain same as previous chapter
public class GameView extends SurfaceView implements SurfaceHolder.Callback
{
    //Same variables as defined earlier

    private ExplosionEffect explosionEffect;
    private long startReset;
    private boolean reset;
    private  boolean started;

    private Random rnd = new Random();
    //GameView constructor, SurfaceChanged and surfaceDestroyed methods
    remain same
    @Override
    public void surfaceCreated(SurfaceHolder holder){
        //bgImg, playerCharacter, rocks, upperBoundary and
        lowerBoundary code same as before

        mainThread = new MainGameThread(getHolder(), this);
        //main thread code after this as earlier

    }
    @Override
    public boolean onTouchEvent(MotionEvent event)
    {
        if(event.getAction() == MotionEvent.ACTION_DOWN) {
            if(!playerCharacter.getPlaying()
            && newGameCreated && reset){
                playerCharacter.setPlaying(true);
                playerCharacter.setUp(true);
            }
            if(playerCharacter.getPlaying()){
```

```
                if(!started)started = true;
                reset = false;
                playerCharacter.setUp(true);
            }
            return true;
        }
    //MotionEvent.ACTION_UP code same as earlier

    return super.onTouchEvent(event);
}

public void update()
{
    if(playerCharacter.getPlaying()) {
        //Same code as earlier
    } else {
        playerCharacter.resetDYC();
        if(!reset) {
            newGameCreated = false;
            startReset = System.nanoTime();
            reset = true;
            explosionEffect = new ExplosionEffect(BitmapFactory
            .decodeResource(getResources(),R.drawable.explosion),
            playerCharacter.getXC(),playerCharacter.getYC()
            -30,100,100,25);
        }

        explosionEffect.update();
        long resetElapsed = (System.nanoTime()-startReset)/1000000;

        if(resetElapsed > 2500 && !newGameCreated) {
            newGame();
        }

        if(!newGameCreated) {
            newGame();
        }
    }
}

//collision code remains same. no change

@Override
public void draw(Canvas canvas)
{
        //Same as till lower boundary and upper boundary code block
        if(started) {
            explosionEffect.draw(canvas);
```

```
            }

            canvas.restoreToCount(savedState);
        }
    }

    //No change in updateUpperBound, updateLowerBound and update method
}
```

So, now you can test your explosion in the game by playing it on your device!:

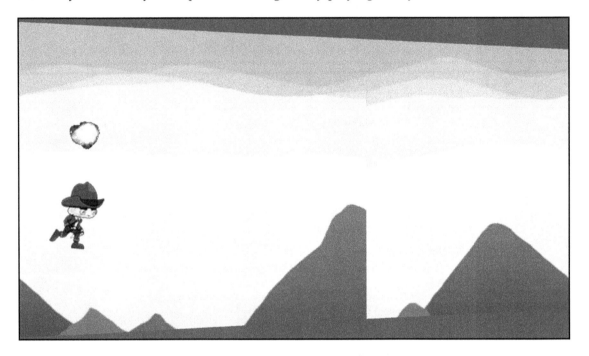

Our explosion effect in action

Kaboom! We're now done with our sprite explosion effect! Now, we will move on to the final part of this game-- the UI.

Creating the UI for our game

If you are not familiar with the term the UI, you must be wondering what the heck is it? UI is an abbreviation for User Interface. To put simply, a UI can consist of all the information you need to display on your game screen or your onscreen controls. Common elements of a UI include the following:

- Text displayed on the screen
- Buttons
- Joystick pad
- Tutorial instructions

In this part of our chapter, we will learn how to display text on the screen. We will also instruct the player how to play the game. We will display the following on our screen:

- Distance ran
- Best score
- Instructions on how to play

So, here we have to display the best score. However, we haven't created our best score variable yet. For this part, we will work entirely in our `GameView.java` file. So, let's define our best score variable in this class:

```
private int bestScore;
```

We're now ready to show our UI data on the screen. Our UI will be entirely based on the `draw()` function, and so we will define a method called as `drawText()`, which should be called from our `draw()` method in the class. So, before we actually call our `drawText()` method, let's write some code for it.

Before that, let's calculate our best score first. Now, obviously, our best score will be calculated after our first game is over, so we will put that logic into our `newGame()` function. The logic for this is fairly simple. If our current score is greater than the `bestScore` that is initialized to 0, then our `bestScore` equals our current score that we get through `playerCharacter.getScore();` and that would be in our `newGame()` function:

```
if(playerCharacter.getScore() > bestScore) {
    bestScore = playerCharacter.getScore();
}
```

Alright, that's sorted, and now we have our bestScore variable ready to hold our best score; our newGame() function would look like this:

```
public void newGame () {
    //clear code and minBoundaryHeight, maxBoundaryHeight code same as
    before
    if(playerCharacter.getScore() > bestScore) {
        bestScore = playerCharacter.getScore();
    }
    //Rest of the code block after this is the same as previous
}
```

We can now write our drawText() method. To do so, we will use the help of the Paint class in Android. The Paint class holds the information about style; color, and how to draw geometries, texts, and bitmaps. Using this class, we can define the color, size, and typeface of our text. Then using our canvas as a reference, we can draw text on our canvas. So let's display our current distance and best score on the screen:

```
public void drawText(Canvas canvas) {
    Paint p = new Paint();
    p.setColor(Color.BLACK);
    p.setTextSize(30);
    p.setTypeface(Typeface.create(Typeface.DEFAULT, Typeface.BOLD));
    canvas.drawText("DISTANCE: "+
    (playerCharacter.getScore()*3),10,HEIGHT-10,p);
    canvas.drawText("BEST: "+ bestScore,WIDTH - 215,HEIGHT-10,p);
}
```

Okay, that looks awesome, but hey, we're still left with one part: the tutorial. As soon as the game starts, we need to instruct the player how to play the game. So, we will add an if statement to control the visibility of our tutorial message. We will guide our player to Tap To Start, Keep Pressed To Go Up, and Release To Go Down:

```
if(!playerCharacter.getPlaying() && newGameCreated && reset) {

    Paint p1 = new Paint();
    p1.setTextSize(40);
    p1.setTypeface(Typeface.create(Typeface.DEFAULT, Typeface.BOLD));
    canvas.drawText("TAP TO START",WIDTH/2-50,HEIGHT/2,p1);

    p1.setTextSize(20);
    canvas.drawText("KEEP PRESSED TO GO UP",WIDTH/2 -
    50,HEIGHT/2+20,p1);
    canvas.drawText("RELEASE TO GO DOWN",WIDTH/2 - 50,HEIGHT/2+40,p1);
}
```

Alright, we are done with our `drawText()` method, and in its entirety it would look like this:

```
public void drawText(Canvas canvas) {
    Paint p = new Paint();
    p.setColor(Color.BLACK);
    p.setTextSize(30);
    p.setTypeface(Typeface.create(Typeface.DEFAULT, Typeface.BOLD));
    canvas.drawText("DISTANCE: "+
    (playerCharacter.getScore()*3),10,HEIGHT-10,p);
    canvas.drawText("BEST: "+ bestScore,WIDTH - 215,HEIGHT-10,p);

    if(!playerCharacter.getPlaying() && newGameCreated && reset) {

        Paint p1 = new Paint();
        p1.setTextSize(40);
        p1.setTypeface(Typeface.create(Typeface.DEFAULT,
        Typeface.BOLD));
        canvas.drawText("TAP TO START",WIDTH/2-50,HEIGHT/2,p1);

        p1.setTextSize(20);
        canvas.drawText("KEEP PRESSED TO GO UP",WIDTH/2 -
        50,HEIGHT/2+20,p1);
        canvas.drawText("RELEASE TO GO DOWN",WIDTH/2 -
        50,HEIGHT/2+40,p1);
    }
}
```

Now, there's one last thing remaining to do. We need to call our `drawText()` method. Any guesses where we will do this? We will do so in our `draw()` method of our class; let's do it:

```
@Override
public void draw(Canvas canvas)
{
    final float scaleFactorX = getWidth()/WIDTH;
    final float scaleFactorY = getHeight()/HEIGHT;
    if(canvas!=null) {
        //No changes in code till if(started) statement

        if(started) {
            explosionEffect.draw(canvas);
        }

        drawText(canvas);

        canvas.restoreToCount(savedState);
    }
}
```

You are all set with your code for displaying the text components on our screen and now have a working game with a UI to display your tutorial message, distance covered, and best score on the screen.

Let's review our code changes that we did in this part of our chapter; code changes are marked in bold:

```
//package name and import statements remain the same as before
public class GameView extends SurfaceView implements SurfaceHolder.Callback
{
    //no change in variable names
    //GameView constructor. No change needed, write as is
    //surfaceChanged method constant as before
    //surfaceDestroyed method same as before
    //surfaceCreated method same as before

    //onTouchEvent same as before
    //collision method written as is

    @Override
    public void draw(Canvas canvas)
    {
        final float scaleFactorX = getWidth()/WIDTH;
        final float scaleFactorY = getHeight()/HEIGHT;
        if(canvas!=null) {
            final int savedState = canvas.save();
            canvas.scale(scaleFactorX, scaleFactorY);
            bgImg.draw(canvas);
            playerCharacter.draw(canvas);

            for(Rock r : rocks) {
                r.draw(canvas);
            }

            for(UpperBoundary ub : upperBoundary){
                ub.draw(canvas);
            }

            for(LowerBoundary lb: lowerBoundary) {
                lb.draw(canvas);
            }

            if(started) {
                explosionEffect.draw(canvas);
            }

            drawText(canvas);
```

```
            canvas.restoreToCount(savedState);
        }
    }
//updateUpperBound code remains same
//updateLowerBound code remains same

public void newGame () {
    lowerBoundary.clear();
    upperBoundary.clear();
    rocks.clear();

    minBoundaryHeight = 5;
    maxBoundaryHeight = 30;

    if(playerCharacter.getScore() > bestScore) {
        bestScore = playerCharacter.getScore();
    }

    playerCharacter.resetScore();
    playerCharacter.resetDYC();
    playerCharacter.setYC(HEIGHT/2);


    for(int i=0; i*20<WIDTH+40;i++) {
        if(i==0) {
            upperBoundary.add(new UpperBoundary
            (BitmapFactory.decodeResource(getResources(),
            R.drawable.ground),i*20,0,10));
        } else {
            upperBoundary.add(new UpperBoundary
            (BitmapFactory.decodeResource(getResources(),
            R.drawable.ground),i*20,0,
            upperBoundary.get(i-1).getHeight()+1));
        }
    }

    for(int i = 0; i*20<WIDTH+40;i++) {
        if(i==0) {
            lowerBoundary.add(new LowerBoundary
            (BitmapFactory.decodeResource(getResources(),
            R.drawable.ground),i*20, HEIGHT-minBoundaryHeight));
        } else {
            lowerBoundary.add(new LowerBoundary
            (BitmapFactory.decodeResource
            (getResources(),R.drawable.ground),
            i*20, lowerBoundary.get(i-1).getYC()-1));
        }
    }
```

```
            newGameCreated = true;
    }

public void drawText(Canvas canvas) {
        Paint p = new Paint();
        p.setColor(Color.BLACK);
        p.setTextSize(30);
        p.setTypeface(Typeface.create(Typeface.DEFAULT,
        Typeface.BOLD));
        canvas.drawText("DISTANCE: "+
        (playerCharacter.getScore()*3),10,HEIGHT-10,p);
        canvas.drawText("BEST: "+ bestScore,WIDTH - 215,HEIGHT-10,p);

        if(!playerCharacter.getPlaying() && newGameCreated && reset) {

            Paint p1 = new Paint();
            p1.setTextSize(40);
            p1.setTypeface(Typeface.create(Typeface.DEFAULT,
            Typeface.BOLD));
            canvas.drawText("TAP TO START",WIDTH/2-50,HEIGHT/2,p1);

            p1.setTextSize(20);
            canvas.drawText("KEEP PRESSED TO GO UP",WIDTH/2 -
            50,HEIGHT/2+20,p1);
            canvas.drawText("RELEASE TO GO DOWN",WIDTH/2 -
            50,HEIGHT/2+40,p1);
        }
    }

    }
```

If you have completed all these steps, then you can go ahead and test your game on your device or emulator now. You will get an output like this:

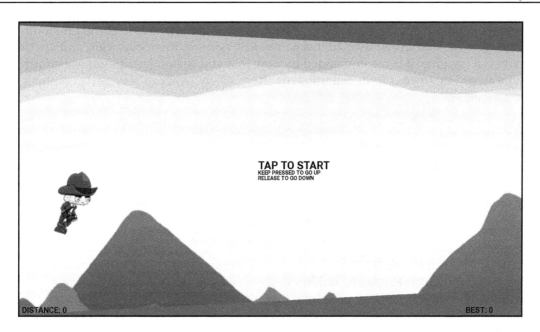

Tutorial displaying our message

We can also see our score in the game while playing:

Distance covered and best score updating as we play

So, here we wrap up our 2D game. Now, based on your understanding of the previous chapters, you can add in elements to this game such as coins, more obstacles, and anything that you can imagine. It is recommended by now that you customize this game as much as possible according to your understanding, or alternatively, you can also start creating a new game altogether.

Summary

We created our explosion and displayed our tutorial message, distance covered, and best score on the screen.

We learned how to create texts on the screen that helped us to display our score, and finally implemented our explosion logic to make the explosion appear on the screen after colliding with rocks. This is it for our 3D game. After this, you can build your game, test it on your device, or even tweak it further to add in more obstacles or make it as interesting as you want. You are only restricted by your imagination.

With this, we wrap up our 2D game, and we will take a look at how to make a shift into the 3D world in our next chapter.

9
Converting Your Game from 2D to 3D

Now, be aware that making a 3D game from scratch in Android Studio is a Herculean task. It isn't absolutely essential to make a game from scratch in 3D because there are plenty of tools and game engines available on the market, which eliminate the requirement of having to work on creating everything from scratch. However, an understanding of how the nuts and bolts operate behind the scenes will definitely help you in the long run.

We will learn the basics of some advanced concepts in this chapter, and by the end of this chapter, you will have an understanding of how to create a basic 3D object in Android Studio. For the purpose of this chapter, we will create a new Android Studio project since we will not continue in our previous game project folder. However, before we get started, let's take a look at the concepts that we will be learning in brief. We will learn the following concepts in this chapter:

- Introduction to OpenGL ES
- Learning about the 3D coordinate system using OpenGL ES
- Creating a scene using OpenGL ES
  - Project folder
  - Render class and main activity
  - Defining shapes

To create our 3D graphics, we will use OpenGL ES for rendering our graphics. Let's understand what OpenGL ES is.

Introduction to OpenGL ES

High-performance 2D and 3D graphics are supported on Android with the OpenGL library. We have the OpenGL ES API, which is used for this. This API specifies the interface for 3D graphics processing hardware. OpenGL is a huge library and from that, a part of it is OpenGL ES which is specifically created for embedded devices.

The various versions of the OpenGL ES API supported on Android are as follows:

- 1.0 and 1.1: Supported by Android 1.0 and higher
- 2.0: Supported by Android 2.2 (API level 8) and higher
- 3.0: Supported by Android 4.3 (API level 18) and higher
- 3.1: Supported by Android 5.0 (API level 21) and higher

You can read more general information about this on the official website for Android here at `https://developer.android.com/guide/topics/graphics/opengl.html`.

Using OpenGL ES, you can create 3D graphics on Android. Since we will be creating 3D objects, we will need to understand the coordinate system. Let's take a quick look at the 3D coordinates system that we will use for our examples.

Learning about the 3D coordinate system

While making our 2D game, we were only dealing with x and y axis values. However, when you are making a 3D game, you have to deal with three axis values: x, y, and z. We're pretty much clear how our x and y axis values work by now. In a similar way, our z axis is projected on the front and back of our mobile device. The following image will explain the three axes in our 3D coordinate system better:

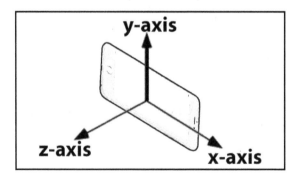

The positive x, y, and z axis on phone

The axes you see in the preceding image are positive directions. If you take their opposite sides, you will get negative values. The origin point starts from (0,0,0), and your values can be float values that will denote your object's location in 3D space.

Here's a classic problem that you might encounter while dealing with OpenGL in terms of device screen sizes. The grid in OpenGL assumes a screen, which is square and has a uniform coordinate system. However, if your screen size varies, then a non-square screen is considered as if it is a perfectly square screen. In order to understand this, take a look at the following figure:

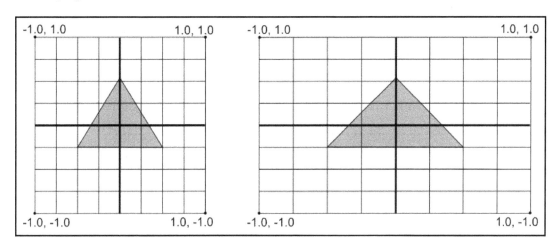

Understanding the default OpenGL coordinate system

As you can see in the preceding figure, we have two different screens, one is square shaped and the other is a rectangle. Now, what happens here is that even though your screen is a rectangle, it is considered a square and because of that, your graphics get stretched. To tackle this issue, you need to apply OpenGL projection modes and camera views to transform the coordinates in order to get a proper proportion on any display.

To do so, we create a projection matrix and a camera view matrix, and then apply that to our OpenGL rendering pipeline. The projection matrix helps us to process the coordinates by recalculating the matrix of our graphics so that they are mapped correctly onto our device screens. The camera view matrix helps create a transformation that renders the object from a specific eye position.

You can read up more about OpenGL ES over at https://en.wikipedia.org/wiki/OpenGL _ES.

Now that we have learned about these basic concepts, we can get started with creating our 3D game scene using OpenGL ES in Android; let's begin!

Creating a 3D scene with OpenGL ES

Just as we did in our 2D game example, most of our draw mechanism is going to remain the same. You will see that there are many similarities between both procedures. We will be working on a new project folder for this example, so let's go ahead and create a new project as we did in `Chapter 8`, *Adding Explosion and Creating a UI*.

Creating our project folder

We will follow the steps we did in `Chapter 2`, *Getting Familiar with Android Studio*. First, we will create a new project from our top menu.

To create a new project, perform the following steps:

1. Go to **File** | **New** | **New Project...**, as shown in the following screenshot:

2. Fill in your **Application name**, **Company domain**, and **Package name** details; then click on **Next**:

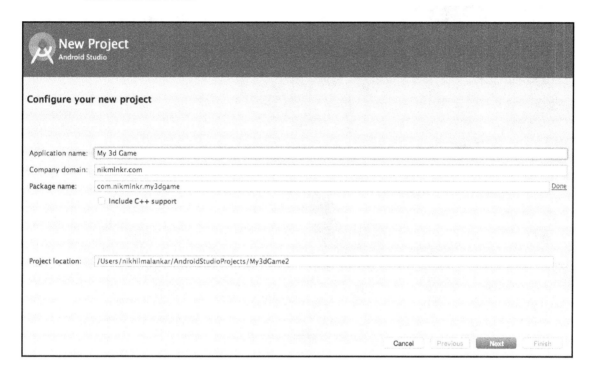

3. Select your target devices and click on **Next**:

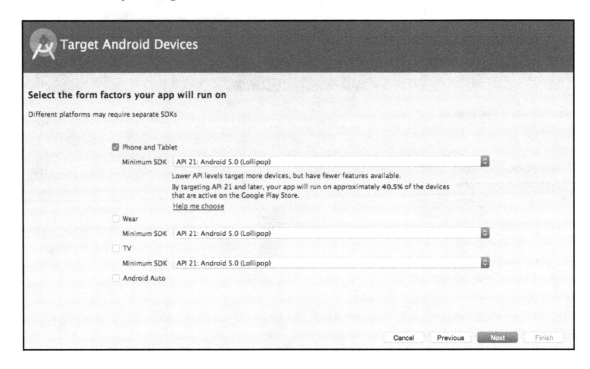

4. Select **Empty Activity** and click on **Next**:

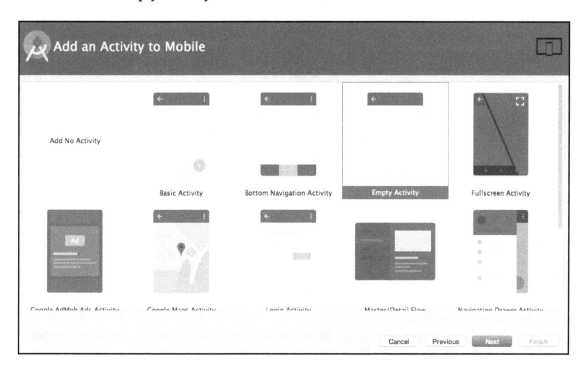

5. Fill in your **Activity Name**, **Layout Name**, and click on **Finish**:

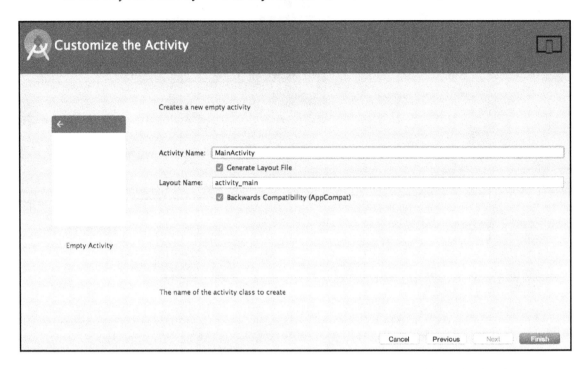

Okay then, we are now set with our new project folder. Also, we will use the landscape mode orientation for this project, so we will define it in our `AndroidManifest.xml` file. Open up your manifest file from the `app/manifests/AndroidManifest.xml` file and make the following change that is marked in bold:

```xml
<?xml version="1.0" encoding="utf-8"?>
<manifest xmlns:android="http://schemas.android.com/apk/res/android"
    package="com.nikmlnkr.my3Dgame">

    <application
        android:allowBackup="true"
        android:icon="@mipmap/ic_launcher"
        android:label="@string/app_name"
        android:roundIcon="@mipmap/ic_launcher_round"
        android:supportsRtl="true"
        android:theme="@style/AppTheme">
    <activity android:name=".MainActivity"
            android:screenOrientation="landscape">
            <intent-filter>
                <action android:name="android.intent.action.MAIN" />
```

```
                    <category android:name="android.intent.category.LAUNCHER"
/>
            </intent-filter>
        </activity>
    </application>

</manifest>
```

Let's now define our main activity and create a render class for our 3D game view.

Creating a render class and defining a main activity

In order to see anything on the screen, we have to render it on screen. Therefore, we need to create a render class that will handle the rendering part of our functioning. Rendering is the process of generating an image from a 2D or 3D model by means of computer programs. We will name our render class as `MyGLRenderer`, but before we do that, let's define our main activity. So, open up your `MainActivity.java` file, and we will create three default methods there. Remove everything you see on screen, except for the first line that's your package name. We will have three methods:

- `onCreate()`: This initializes our main activity
- `onPause()`: This handles the app if it goes into background
- `onResume()`: After the user resumes back to the app, this method is called

Also, we are dealing with OpenGL graphics here, and for that reason, we will need a `GLSurfaceView`, just as we worked in our 2D game wherein we had a surface view for our game.

So, let's create them in our `MainActivity.java` file, as follows:

```
package com.nikmlnkr.my3Dgame;

import android.app.Activity;
import android.opengl.GLSurfaceView;
import android.os.Bundle;

public class MainActivity extends Activity {

    private GLSurfaceView gv;

    //Our onCreate method
    @Override
```

```
    protected void onCreate(Bundle savedInstanceState) {
        super.onCreate(savedInstanceState);
        gv = new GLSurfaceView(this);
        gv.setRenderer(new MyGLRenderer(this));
        this.setContentView(gv);
    }

    //Resume method
    @Override
    protected void onResume() {
        super.onResume();
        gv.onResume();
    }

    //Pause method
    @Override
    protected void onPause() {
    super.onPause();
        gv.onPause();
    }
}
```

Okay, that seems sorted, but now we are getting an error on our `MyGLRenderer` line. That's because we still haven't made our renderer class yet. Let's make our renderer class now. Create a new Java class file named `MyGLRenderer.java`, and let's get started with writing our renderer class.

 The `Renderer` class is where you will do most of your object displaying part.

The interface `GLSurfaceView.Renderer` is responsible for making OpenGL render a frame, hence we will need to implement it as an interface in our code; so we will start with our very first line, which will do the same. We will extend our class to the `GLSurfaceView.Renderer` interface and write a default constructor, as follows:

```
public class MyGLRenderer implements GLSurfaceView.Renderer {
    Context ct;      //Context variable

    //Constructor of our renderer class
    public MyGLRenderer(Context ct) {
        this.ct = ct;
    }
}
```

Now, inside this class, as we had in our 2D sprite example game, we will need three methods to draw on the screen; they are `onSurfaceCreated()`, `onSurfaceChanged()`, and `onDraw()`. However, in OpenGL, the `onDraw()` method is actually `onDrawFrame()`, so let's define those three methods one by one.

First, we will start with our `onSurfaceChanged()` method. In this method, we will create our surface. This method is for the initialization of our scene. Here, we will create a simple black screen drawn on the screen. In order to set our color, we will use RGBA values. For a black color, we will need (0,0,0,1) RGBA values. We will also add in some more initialization factors here:

```
@Override
public void onSurfaceCreated(GL10 gles, EGLConfig c) {
    gles.glClearColor(0.0f, 0.0f, 0.0f, 1.0f);
    //Clear color and set to black
    gles.glClearDepthf(1.0f);
    //Clear depth
    gles.glEnable(GL10.GL_DEPTH_TEST);
    //Enable depth test
    gles.glDepthFunc(GL10.GL_LEQUAL);
    //Set depth function

    gles.glHint(GL10.GL_PERSPECTIVE_CORRECTION_HINT,
    GL10.GL_NICEST);
    //set gl to nicest

    gles.glShadeModel(GL10.GL_SMOOTH);
    //set shade model to smooth
    gles.glDisable(GL10.GL_DITHER);
    //disable dither
}
```

These parameters are basically required to set various aspects in an OpenGL ES environment. You can read up more about these specific methods and more through `https://www.khronos.org/registry/OpenGL-Refpages/es3.0/`.

This method gets called after the `onSurfaceCreated()` method and also every time our screen resolution changes. Basically, this method is responsible for creating the display matrix that we saw earlier in the chapter, which in turn creates a uniform shape on any screen:

```
@Override
public void onSurfaceChanged(GL10 gles, int w, int h) {
    if (h == 0) h = 1;
    float aspect = (float)w / h;
```

```
        gles.glViewport(0, 0, w, h);
        //dynamically set the width and height of our viewport as per
screen
        resolution

        gles.glMatrixMode(GL10.GL_PROJECTION);
        //set our matrix mode projection
        gles.glLoadIdentity();
        GLU.gluPerspective(gles, 45, aspect, 0.1f, 100.f);

        gles.glMatrixMode(GL10.GL_MODELVIEW);
        //set our camera view matrix mode

        gles.glLoadIdentity();
    }
```

Finally, we have our `onDrawFrame()` function, which is used to draw the current frame. After every frame, we need to clear the previous screen drawn, and for that reason, we call `glClear()` in this before we render any code further. Right now, we will just write our clear code and when we will draw our object shapes in this function after our clear code:

```
@Override
public void onDrawFrame(GL10 gles) {
    gles.glClear(GL10.GL_COLOR_BUFFER_BIT |
    GL10.GL_DEPTH_BUFFER_BIT);
    //clear depth buffer
}
```

Finally, this is how your entire code will look like for your `Renderer` class:

```
package com.nikmlnkr.my3Dgame;

/**
 * Created by nikhilmalankar on 05/03/17.
 */
import javax.microedition.khronos.egl.EGLConfig;
import javax.microedition.khronos.opengles.GL10;
import android.content.Context;
import android.opengl.GLSurfaceView;
import android.opengl.GLU;

public class MyGLRenderer implements GLSurfaceView.Renderer {
    Context ct;     //Our context variable

    // Constructor of our renderer
    public MyGLRenderer(Context ct) {
        this.ct = ct;
    }
```

```
// Call back when the surface is first created or re-created
@Override
public void onSurfaceCreated(GL10 gles, EGLConfig c) {
    gles.glClearColor(0.0f, 0.0f, 0.0f, 1.0f);
    gles.glClearDepthf(1.0f);
    gles.glEnable(GL10.GL_DEPTH_TEST);
    gles.glDepthFunc(GL10.GL_LEQUAL);
    gles.glHint(GL10.GL_PERSPECTIVE_CORRECTION_HINT,
    GL10.GL_NICEST);
    gles.glShadeModel(GL10.GL_SMOOTH);
    gles.glDisable(GL10.GL_DITHER);
}

@Override
public void onSurfaceChanged(GL10 gles, int w, int h) {
    if (h == 0) h = 1;
    float aspect = (float)w / h;

    gles.glViewport(0, 0, w, h);

    gles.glMatrixMode(GL10.GL_PROJECTION);

    gles.glLoadIdentity();
    GLU.gluPerspective(gles, 45, aspect, 0.1f, 100.f);

    gles.glMatrixMode(GL10.GL_MODELVIEW);
    gles.glLoadIdentity();
}

@Override
public void onDrawFrame(GL10 gles) {
    gles.glClear(GL10.GL_COLOR_BUFFER_BIT |
    GL10.GL_DEPTH_BUFFER_BIT);
    //clear our depth buffer
}
}
```

Now, we will learn how to define and draw, or in proper terms *render*, our 3D objects on the screen. First, we will start with a basic triangle shape and then we will draw a pyramid, then we will make both of them rotate on the screen.

Defining shapes

In order to draw something on the screen, we first need to define its shape and then render it. So we won't define our shape in our render class. We will create a new class in order to define our shape. As discussed, we will first create a basic triangle.

In order to create our triangle, we need to first define its vertices. So, let's create a new class named `Triangle.java` and start writing our logic to define the shape of our 3D triangle. Before we start writing our code, let's try to understand what we are actually going to do.

We will take vertices and plot them on our screen. Also, as you saw previously how OpenGL coordinates work, we will plot our vertices in all three directions. For our triangle, we require three points to be plotted. So, we will plot one point on the positive **y-axis** and the other two points on the positive and negative x-axis:

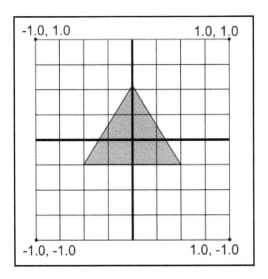

Plotting our triangle

For this purpose, we will take a three-dimensional array of our vertices and define them as a variable in our class:

```
package com.nikmlnkr.my3Dgame;

/**
 * Created by nikhilmalankar on 05/03/17.
 */
public class Triangle {
    private float[] v = {  // Vertices of our triangle
        0.0f,  1.0f, 0.0f, // 0. top vertices
```

```
            -1.0f, -1.0f, 0.0f, // 1. left-bottom vertices
            1.0f, -1.0f, 0.0f  // 2. right-bottom vertices
        };
    }
```

That takes care of our vertices, and we have defined its shape, well somewhat, but of course, there's more to it. We also need to define a vertex buffer and transfer this data into it. To do so, we will define our vertex buffer variable as nio's buffer, as they are placed on the native heap and are not garbage collected. We will do the same for our index buffer, which will arrange our triangle in a counter-clockwise (CCW) direction with positive *z* direction facing toward the screen. So, first we will define our vertexBuffer and indexBuffer variables, and then in our default constructor, we will set up our vertexBuffer and indexBuffer:

```
        private FloatBuffer vb;
        private ByteBuffer ib

        private byte[] ind = { 0, 1, 2 };

        public Triangle() {
            ByteBuffer vbb = ByteBuffer.allocateDirect(v.length * 4);
            vbb.order(ByteOrder.nativeOrder());
            vb = vbb.asFloatBuffer();
            vb.put(v);
            vb.position(0);

            ib = ByteBuffer.allocateDirect(ind.length);
            ib.put(ind);
            ib.position(0);
        }
```

Alright, now that's taken care of, we also have to actually draw our triangle on screen; for this, we will use our draw() method. To do so, we will go through the following four simple steps:

1. We enable vertex array client states.
2. We specify the location of the buffers.
3. We render our primitive shapes using glDrayElements() that uses index array to reference the vertex array.
4. We disable our vertex array client state.

Now that we have understood the working of its theory, let's go ahead and do this in practice by writing it out in our code:

```java
public void draw(GL10 gles) {
    gles.glEnableClientState(GL10.GL_VERTEX_ARRAY);
    gles.glVertexPointer(3, GL10.GL_FLOAT, 0, vb);

    gles.glDrawElements(GL10.GL_TRIANGLES, ind.length,
    GL10.GL_UNSIGNED_BYTE, ib);
    gles.glDisableClientState(GL10.GL_VERTEX_ARRAY);
}
```

Our shape is defined perfectly, and our entire code block for the `Triangle.java` file will look like this:

```java
package com.nikmlnkr.my3Dgame;

/**
 * Created by nikhilmalankar on 05/03/17.
 */

import java.nio.ByteBuffer;
import java.nio.ByteOrder;
import java.nio.FloatBuffer;
import javax.microedition.khronos.opengles.GL10;

public class Triangle {
    private FloatBuffer vb;
    private ByteBuffer ib;

    private float[] v = {  // Vertices of our triangle
        0.0f,  1.0f, 0.0f, // 0. top vertices
        -1.0f, -1.0f, 0.0f, // 1. left-bottom vertices
        1.0f, -1.0f, 0.0f  // 2. right-bottom vertices
    };
    private byte[] indices = { 0, 1, 2 };

    public Triangle() {
        ByteBuffer vbb = ByteBuffer.allocateDirect(v.length * 4);
        vbb.order(ByteOrder.nativeOrder());
        vb = vbb.asFloatBuffer();
        vb.put(v);
        vb.position(0);

        ib = ByteBuffer.allocateDirect(ind.length);
        ib.put(ind);
        ib.position(0);
    }
```

```
public void draw(GL10 gles) {
    gles.glEnableClientState(GL10.GL_VERTEX_ARRAY);
    gles.glVertexPointer(3, GL10.GL_FLOAT, 0, vb);

    gles.glDrawElements(GL10.GL_TRIANGLES, ind.length,
    GL10.GL_UNSIGNED_BYTE, ib);
    gles.glDisableClientState(GL10.GL_VERTEX_ARRAY);
}
}
```

However, if you deploy this to your device, you won't be able to see your triangle on the screen. This is because we still haven't rendered it on our screen yet. Remember how we discussed rendering all our objects using the help of the MyGLRenderer file? We haven't told that file to use our triangle to be rendered, but that's it for this chapter. We will cover the rendering of our object in the next chapter.

Summary

In this chapter, we learned about how to transition from a 2D to a 3D game along with concepts of OpenGL. We also learned how to create our main activity along with our own renderer. After this, we learned how to define a basic shape.

To summarize, we covered the following topics in this chapter. We got introduced to OpenGL ES and took a look at the coordinates system in the same. We learned how to create a blank scene and the process of rendering. We then created our basic renderer and defined our triangle shape.

In our next chapter, we will learn how to render our defined shape on the screen and rotate it. We will also learn how to create a 3D pyramid in the next chapter, and by the end of the next chapter, you will complete this book and also have a foundation for creating a 2D and 3D game.

10
Working Further on 3D Game

Now that we have defined our 3D shape, let's render it on our screen. However, before we begin this chapter, let's summarize everything we learned throughout the contents of this book since this is our last chapter.

We learned a lot of cool stuff about Android. We started from learning about the history of Android and quickly progressed toward the installation of software required to develop apps for Android. After that, we installed Android Studio and configured it with the latest components of Android-N.

Once our environment was set up, we learned how to create a basic app for Android and learned about various concepts of Android, such as package names, inputs, emulators, and more. After getting an air of familiarity with Android, we learned about how to make a transition from making apps to games wherein we learned how to create a surface and canvas, which would allow us to create graphics for our game, and then we moved on to learn various concepts about game development.

As we progressed further through our chapters, we started learning about the process of making games, and within a couple of chapters, we learned how to create a 2D game from scratch with a UI. We completed developing a fully-fledge 2D game with the same and then moved on to turning things into 3D.

In our previous chapter, we learned that we can create 3D games for native Android using OpenGL ES, and so we started with defining basic shapes.

In this last chapter for this book, we will learn the following topics:

- Rendering our object on screen
- Adding color to our object
- Rotating both of our 3D objects
- Creating a pyramid

So without further ado, let's begin with this final part of our book, which will set us up with the foundation needed to begin your journey in 3D game development.

Rendering our object on screen

In our previous chapter, we created a class for our `Triangle` object; however, if you run your game, it will still show a blank screen since we have not used our render class to display it. We need to create an object for our newly defined class, and then using our GL reference, we will draw/render it on our screen. Open up your `MyGLRenderer.java` file and let's start by declaring a variable of our triangle. We won't be altering any of our code that we wrote in `onSurfaceCreated()` or `onSurfaceChanged()` methods.

In order to draw our object on screen, we will simply follow these steps:

1. Define a variable of our `Triangle` class.
2. Assign a reference to it in our constructor.
3. Using gl, access the `draw()` method in our `Triangle` to display it on screen.

Let's take a look at how we can do that; just type in the code marked in bold in your `MyGLRenderer.java` file:

```
//The import statements are same as our previous chapter

public class MyGLRenderer implements GLSurfaceView.Renderer {

    Triangle triangle;

    // Constructor
    public MyGLRenderer(Context context) {
        triangle = new Triangle();
    }

    //onSurfaceCreated and onSurfaceChanged methods remain same as previous
chapter
```

```
@Override
public void onDrawFrame(GL10 gles) {
    gles.glClear(GL10.GL_COLOR_BUFFER_BIT |
    GL10.GL_DEPTH_BUFFER_BIT);

    gles.glLoadIdentity();
    gles.glTranslatef(-1.5f, 0.0f, -6.0f);
    triangle.draw(gles);
    }
}
```

Pretty simple, right? That's it! Your triangle object is ready to be rendered on screen. Don't believe it? Go ahead; test and run it on your emulator/device, and you will get the following output:

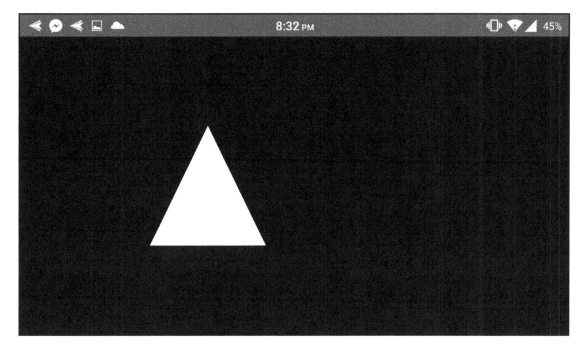

Our triangle rendered on screen, finally!

You might think that this object is still 2D, but wait till the part where we rotate it. However, before that, let's learn how to add color to this object. So let's add some colors.

Adding colors to our object

In this part, we will demonstrate how we can use different colors to texture our object. We will use RGB values using a `colorBuffer` to fetch values from the colors of the vertices. After this, we will enable the color-array client state, and then these colors are rendered together with the vertices in `glDrawElements()`.

Here, we will again use nio's `FloatBuffer` to declare our `colorBuffer` variable. Here are the steps we would use to add color to our object:

1. We declare our `colorBuffer` variable.
2. We declare our color array variable.
3. We copy our color vertices data to our buffer.
4. Enable our color array.
5. Define color array buffer.
6. Disable color array.

Also, since this is a native property of the triangle, we will write this code in our `Triangle.java` class that we created and not in the `MyGLRenderer.java` class. Type in the code marked in bold keeping in mind the preceding steps:

```java
//Import statements remain the same
public class Triangle {
private FloatBuffer vb;
    private FloatBuffer cb;
    private ByteBuffer ib;

    private float[] v = {  // Vertices of the triangle
            0.0f,  1.0f, 0.0f, // 0. top vertices
            -1.0f, -1.0f, 0.0f, // 1. left-bottom vertices
            1.0f, -1.0f, 0.0f  // 2. right-bottom vertices
    };
    private byte[] ind = { 0, 1, 2 };
    private float[] colors = {
            1.0f, 0.0f, 0.0f, 1.0f, //R
            0.0f, 1.0f, 0.0f, 1.0f, //G
            0.0f, 0.0f, 1.0f, 1.0f  //B
    };

    public Triangle() {
        //start of code same as earlier

        ByteBuffer cbb = ByteBuffer.allocateDirect(colors.length * 4);
        cbb.order(ByteOrder.nativeOrder());
```

```
        cb = cbb.asFloatBuffer();
        cb.put(colors);
        cb.position(0);

        //index buffer code same as earlier
    }

    public void draw(GL10 gles) {
        //code as before
        gles.glDisableClientState(GL10.GL_COLOR_ARRAY);
    }
}
```

Now, we have a colored triangle. You will get an output like this if you used the specified color values. Of course, you are free to change and tweak the colors as you please by changing the RGB values in our colors variable:

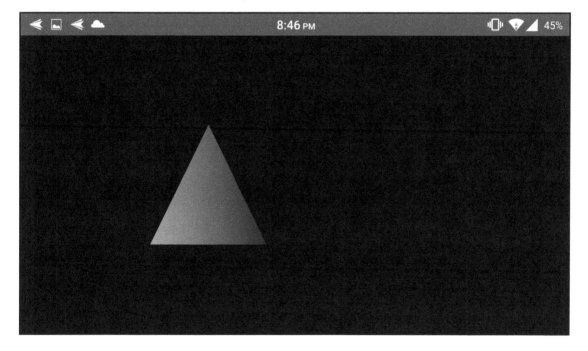

Our colored triangle

Okay, now we will make our triangle rotate, which will clarify our doubts whether it is actually a 3D object or not.

Rotating our object

By now, you must have understood that any changes in the rendering of an object have to be done in the `MyGLRenderer`, whereas any properties that are local to our object are to be done in the respective object's file. So, we will write our rotation code in our `MyGLRenderer.java` file because rotating an object is a part of the rendering process.

Here's our logic to rotate our triangle:

- We take our rotation angle
- We rotate our object in a specified rotation angle
- We increment our rotation angle

Let's do it; we will simply declare two variables at the start for our rotational angle and speed; then in our `onDrawFrame()` method, implement our rotation logic, as follows:

```
//Import statements as before
public class MyGLRenderer implements GLSurfaceView.Renderer {

    Triangle triangle;

    private float angleTriangle = 0.0f;
    private float speedTriangle = 0.5f;

    public MyGLRenderer(Context context) {
        triangle = new Triangle();
    }

// No changes in our onSurfaceCreated and onSurfaceChanged methods so
ignore this part

    @Override
    public void onDrawFrame(GL10 gles) {
        //Same as previous part

        angleTriangle += speedTriangle;
    }
}
```

After you compile this code, your triangle will start rotating. Cool, right? Build and run and see your rotating triangle in action!:

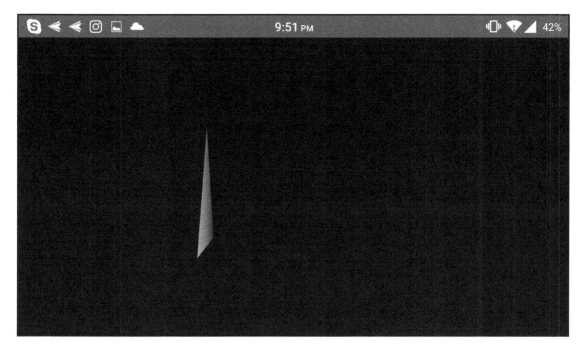

Our rotating triangle

Fantastic! Now, let's use this knowledge to create a proper 3D object, our pyramid! We're pretty clear about the process for creating objects, so we will breeze through the pyramid part.

Let's repeat the same steps for our pyramid. This time our object will be purely 3D rather than just a 2D plane object; so, let's start!

Creating a pyramid 3D object

Now that you have understood how to define shapes and render objects on screen, it will be comparatively easier to create our 3D object. We will follow almost the same procedure as we did to create our triangle. We will create our pyramid right besides our triangle; let's begin.

Defining the shape

As we've learned from the previous part, we will first create a class of our pyramid in order to define our shape. So create a file `Pyramid.java` to define the shape of our pyramid object.

Our pyramid has five faces, so we will require five vertices to draw our pyramid. So after you create your `Pyramid.java`, we will define our vertices, as follows:

```
//Package name of our game

public class Pyramid {
    private float[] vp = { // 5 vertices of the pyramid in (x,y,z)
            -1.0f, -1.0f, -1.0f,   //left-bottom-back
             1.0f, -1.0f, -1.0f,   //right-bottom-back
             1.0f, -1.0f,  1.0f,   //right-bottom-front
            -1.0f, -1.0f,  1.0f,   //left-bottom-front
             0.0f,  1.0f,  0.0f    //top
    };
}
```

Alright, now we have our vertices set, but just like our triangle, we still have our buffers and indices to take care of. We will quickly define our float and byte buffers for our shape and colors along with the indices that will make our pyramid faces:

```
private FloatBuffer vb;   // Buffer for vertex-array
private FloatBuffer cb;   // Buffer for color-array
private ByteBuffer ib;    // Buffer for index-array

private float[] colors = {  // Colors of the 5 vertices in RGBA
        0.0f, 0.0f, 1.0f, 1.0f,   // blue
        0.0f, 1.0f, 0.0f, 1.0f,   // green
        0.0f, 0.0f, 1.0f, 1.0f,   // blue
        0.0f, 1.0f, 0.0f, 1.0f,   // green
        1.0f, 0.0f, 0.0f, 1.0f    // red
};

private byte[] ind = { // Vertex indices
        2, 4, 3,   // front face
        1, 4, 2,   // right face
        0, 4, 1,   // back face
        4, 0, 3    // left face
};
```

Now, we will write the constructor of our `Pyramid` based on the logic of our triangle:

```
public Pyramid() {
    ByteBuffer vbb = ByteBuffer.allocateDirect(vp.length * 4);
    vbb.order(ByteOrder.nativeOrder());
    vb = vbb.asFloatBuffer();
    vb.put(vp);
    vb.position(0);

    ByteBuffer cbb = ByteBuffer.allocateDirect(colors.length * 4);
    cbb.order(ByteOrder.nativeOrder());
    cb = cbb.asFloatBuffer();
    cb.put(colors);
    cb.position(0);

    ib = ByteBuffer.allocateDirect(ind.length);
    ib.put(ind);
    ib.position(0);
}
```

Finally, of course, our `draw()` method as shown in the following code:

```
public void draw(GL10 gles) {
    gles.glFrontFace(GL10.GL_CCW);

    gles.glEnableClientState(GL10.GL_VERTEX_ARRAY);
    gles.glVertexPointer(3, GL10.GL_FLOAT, 0, vb);
    gles.glEnableClientState(GL10.GL_COLOR_ARRAY);
    gles.glColorPointer(4, GL10.GL_FLOAT, 0, cb);

    gles.glDrawElements(GL10.GL_TRIANGLES, ind.length,
    GL10.GL_UNSIGNED_BYTE, ib);

    gl.glDisableClientState(GL10.GL_VERTEX_ARRAY);
    gl.glDisableClientState(GL10.GL_COLOR_ARRAY);
}
```

We're done defining the shape of our `Pyramid`. So, finally your entire code for `Pyramid` will be like this:

```
//Package name and our import statements
public class Pyramid {

//Our 3 buffer variables

    private float[] vp = {
            -1.0f, -1.0f, -1.0f,  //left-bottom-back
            1.0f, -1.0f, -1.0f,   //right-bottom-back
```

```
                    1.0f, -1.0f,  1.0f,  //right-bottom-front
                   -1.0f, -1.0f,  1.0f,  //left-bottom-front
                    0.0f,  1.0f,  0.0f   //top
        };

        private float[] colors = {
                0.0f, 0.0f, 1.0f, 1.0f,  //blue
                0.0f, 1.0f, 0.0f, 1.0f,  //green
                0.0f, 0.0f, 1.0f, 1.0f,  //blue
                0.0f, 1.0f, 0.0f, 1.0f,  //green
                1.0f, 0.0f, 0.0f, 1.0f   //red
        };

        private byte[] ind = { // Vertex indices
                2, 4, 3,   // front face (CCW)
                1, 4, 2,   // right face
                0, 4, 1,   // back face
                4, 0, 3    // left face
        };

        public Pyramid() {
            ByteBuffer vbb = ByteBuffer.allocateDirect(vp.length * 4);
            vbb.order(ByteOrder.nativeOrder());
            vb = vbb.asFloatBuffer();
            vb.put(vp);
            vb.position(0);

            ByteBuffer cbb = ByteBuffer.allocateDirect(colors.length * 4);
            cbb.order(ByteOrder.nativeOrder());
            cb = cbb.asFloatBuffer();
            cb.put(colors);
            cb.position(0);

            ib = ByteBuffer.allocateDirect(ind.length);
            ib.put(ind);
            ib.position(0);
        }

// Draw the shape
    public void draw(GL10 gles) {
        gles.glFrontFace(GL10.GL_CCW);

        gles.glEnableClientState(GL10.GL_VERTEX_ARRAY);
        gles.glVertexPointer(3, GL10.GL_FLOAT, 0, vb);
        gles.glEnableClientState(GL10.GL_COLOR_ARRAY);
        gles.glColorPointer(4, GL10.GL_FLOAT, 0, cb);

        gles.glDrawElements(GL10.GL_TRIANGLES, ind.length,
```

```
        GL10.GL_UNSIGNED_BYTE,ib);

        gles.glDisableClientState(GL10.GL_VERTEX_ARRAY);
        gles.glDisableClientState(GL10.GL_COLOR_ARRAY);
    }
}
```

Okay, that's set but we still won't be getting to see the Pyramid on our screen since we haven't rendered it yet; let's do that now to bring our pyramid on screen.

Rendering our 3D object

Based on our understanding of our previous example working with the triangle, we are pretty clear that in order to render our objects, we have to use our `MyGLRenderer.java` class. Since we are pretty clear on how our 3D rendering works, we will also add in our rotation code in the same code.

Here's one crucial thing to take into account though, we will be drawing our pyramid beside our triangle, so we have to make sure that we don't overlap our pyramid over our triangle. To do so, we will use the code `gl.Translate()` that we will observe as a follow up inside our `onDrawFrame()` code block after we draw our triangle.

We will work on our `MyGLRenderer.java` file, so open that file and write in the code marked in bold; the rest of the part remains the same:

```
//Package name and import statements

public class MyGLRenderer implements GLSurfaceView.Renderer {

    //our triangle object variables
    private Pyramid pyramid;

    private static float anglePyramid = 0;
    private static float speedPyramid = 2.0f;

    public MyGLRenderer(Context context) {
        //our triangle object reference remains same
        pyramid = new Pyramid();
    }

    //Again there's no change in onSurfaceCreated and onSurfaceChanged
methods so type them as is in previous chapter

    @Override
    public void onDrawFrame(GL10 gles) {
```

```
//Triangle code remains same
gles.glLoadIdentity();
gles.glTranslatef(1.5f, 0.0f, -6.0f);
gles.glRotatef(anglePyramid, 0.1f, 1.0f, -0.1f);
pyramid.draw(gles);

//angleTriangle speed assign here
anglePyramid += speedPyramid;
    }
  }
```

Alright, so everything looks set. Build and run your project, and you will get the following output:

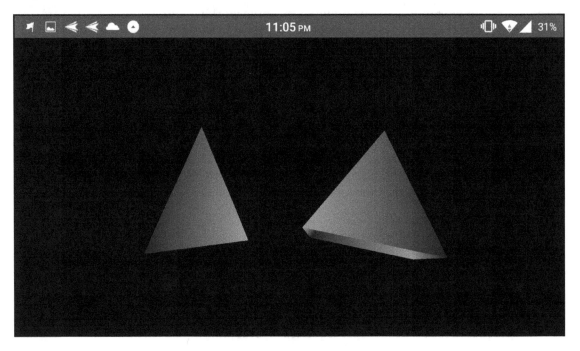

Both our objects are visible on screen now

Congratulations! You have successfully created a 3D object using OpenGL. In a similar way, you can create any kind of object based on the vertices.

This is just the foundation of creating a 3D basic shapes game with OpenGL. Creating a fully-fledged 3D game is a huge task of its own, and this is just a glimpse of what you can do.

We would recommend that you buy our other books on 3D game development to go ahead in more depth on creating games in 3D.

Summary

In this chapter, we learned how to create 3D objects and render them on the screen. We also learned how to add colors to objects as well as rotate them in a 3D direction. With this chapter, we conclude this book.

We have learned how to create games in 2D and 3D using native Android. With the knowledge gained from this chapter, you can get started on your journey into game development and start creating your own games for Android.

This concludes this book for creating your games with Android. The knowledge gained through this book will serve as a foundation for the games you make ahead. The advantage of creating games in native Android is that primarily, the file size of games turns out to be quite small, which is what users generally look for. If you were to create any game using an engine, you would end up creating a game, which would have a comparatively larger file size by the end of the development process; you'd need to optimize it even further, but yet there's only so much you can do for that. So, developing native will not only allow you to make games with smaller file size, but will also enhance your knowledge and core understanding of the entire process.

As we conclude this book, it is now up to you and your imagination to start creating games with the help of the knowledge you learned through this book. Of course, creating games with an engine is definitely a faster process; if you want to skip through creating your own classes for each and everything, it is recommended that you get started with game engines. Packt has a wide variety of books to help you in the process. Perhaps, your best starting point to get started with a game engine is *Unity Game Engine*. If you're interested in learning more and giving your game development process a boost, then the best recommendation will be to go through Packt's book on *Unity Game Engine*.

Wishing you all the best for creating your next game!

Index